Lecture Notes in Computer Science   12039

More information about this series at http://www.springer.com/series/7408

Marco Gribaudo · Mauro Iacono ·
Tuan Phung-Duc · Rostislav Razumchik (Eds.)

Computer Performance Engineering

16th European Workshop, EPEW 2019
Milan, Italy, November 28–29, 2019
Revised Selected Papers

 Springer

Editors
Marco Gribaudo
Politecnico di Milano
Milano, Italy

Tuan Phung-Duc
University of Tsukuba
Tsukuba, Japan

Mauro Iacono
Università degli Studi
della Campania "L. Vanvitelli"
Caserta, Italy

Rostislav Razumchik
Federal Research Center
"Computer Science and Control"
of the Russian Academy of Sciences
Moscow, Russia

ISSN 0302-9743 ISSN 1611-3349 (electronic)
Lecture Notes in Computer Science
ISBN 978-3-030-44410-5 ISBN 978-3-030-44411-2 (eBook)
https://doi.org/10.1007/978-3-030-44411-2

LNCS Sublibrary: SL2 – Programming and Software Engineering

This Springer imprint is published by the registered company Springer Nature Switzerland AG
The registered company address is: Gewerbestrasse 11, 6330 Cham, Switzerland

Preface

Following the tradition of the previous EPEWs, the goal of this workshop was to gather academic and industrial researchers working on all aspects of performance engineering. The papers presented at the workshop were devoted to modeling and analysis of network/control protocols and high performance/big data information systems, analysis of scheduling, blockchain technology, and analytical modeling and simulation of computer/network systems. The call for papers gathered 13 high-quality submissions, and each was peer reviewed by an average of three reviewers from the Program Committee (PC). Each reviewer assessed the relevance, novelty, and technical soundness of the assigned papers. After the reviews were collected, it was decided that 10 papers would be accepted having the highest (also weighted across the reviewers' confidence) score among the positive ones. This year there were two keynote talks given by Prof. Emeritus Giuseppe Serazzi from Politecnico di Milano (Italy) and Prof. Anne Remke from the University of Münster (Germany). Prof. Giuseppe Serazzi addressed current research and recent developments in the JMT software – simulator of SPN and GSPN, CPN and QN models. Prof. Anne Remke explored the state of the art in model checking Hybrid Petri nets, featuring a newly released tool, which performs full-fledged STL model checking efficiently for Petri nets with a finite but arbitrary number of random variables. We thank our keynote speakers, as well as all PC members who placed their reviews on time despite the extremely tight reviewing deadline and provided constructive and insightful comments. We also express our gratitude to the Organizing Committee at Politecnico di Milano for their continuous and valuable help, the EasyChair team for their conference system, and Springer for their continued editorial support. Above all, we would like to thank the authors of the papers for their contribution to this volume, which we hope that you, the reader, will find useful.

February 2020

Marco Gribaudo
Mauro Iacono
Tuan Phung-Duc
Rostislav Razumchik

Organization

General Chairs

Marco Gribaudo   Politecnico di Milano, Italy
Mauro Iacono   Università degli Studi della Campania "Luigi
     Vanvitelli", Italy

Technical Program Chairs

Tuan Phung-Duc   University of Tsukuba, Japan
Rostislav Razumchik   Federal Research Center "Computer Science
     and Control" of the Russian Academy of Sciences,
     Russia

Program Committee

Elvio Gilberto Amparore   University of Turin, Italy
Paolo Ballarini   CentraleSupeléc, France
Enrico Barbierato   Politecnico di Milano, Italy
Marco Bernardo   University of Urbino, Italy
Marko Boon   Eindhoven University of Technology, The Netherlands
Laura Carnevali   University of Florence, Italy
Hind Castel   Télécom SudParis, France
Davide Cerotti   Politecnico di Milano, Italy
Ioannis Dimitriou   University of Patras, Greece
Dieter Fiems   Ghent University, Belgium
Jean-Michel Fourneau   Université de Versailles Saint-Quentin-en-Yvelines,
     France
Marco Gribaudo   Politecnico di Milano, Italy
Boudewijn Haverkort   Tilburg University, The Netherlands
András Horváth   University of Turin, Italy
Esa Hyytiä   University of Iceland, Iceland
Mauro Iacono   Università degli Studi della Campania "Luigi
     Vanvitelli", Italy
Alain Jean-Marie   Inria, France
Lasse Leskelä   Aalto University, Finland
Oleg Lukashenko   Institute of Applied Mathematical Research
     of the Karelian Research Centre of the Russian
     Academy of Sciences, Russia
Andrea Marin   University of Venice, Italy
Marco Paolieri   University of Southern California, USA
Nihal Pekergin   Université Paris-Est Créteil, France

Hybrid Petri Nets Featuring Multiple Random Variables (Keynote)

Anne Remke

Westfälische Wilhelms-Universität, Münster, Germany
anne.remke@uni-muenster.de

Abstract. Hybrid Petri nets have been extended with random variables to model stochastic time delays. This restricted class of stochastic hybrid systems has successfully been used to model critical infrastructures like water sewage systems and smart home energy storage and control. The logic STL has been proposed to formulate properties, which can automatically be model checked for Hybrid Petri nets. Model checking requires to first build the underlying state space and then relies on geometric operations on convex state sets which symbolically represent sets of states with similar properties. After the satisfaction sets are computed a final integration step is needed to compute the probability that a specific STL formula holds. This paper provides an overview on the state-of-the-art in model checking Hybrid Petri nets, featuring a newly released tool, which performs full-fledged STL model checking efficiently for Hybrid Petri nets with a finite but arbitrary number of random variables.

Keywords: Hybrid Petri nets · Model checking · Stochastic hybrid systems

Hybrid Petri nets with general transitions (HPnG) [11] extend Hybrid Petri nets [2] by adding stochastic delays through general transitions which fire after a randomly distributed amount of time. HPnGs provide a high-level and process-oriented formalism for a restricted class of stochastic hybrid systems, where the continuous behaviour is piece-wise linear without resets and the inherent non-determinism is resolved probabilistically. Hybrid Petri nets without the above mentioned stochastic extension form a subclass of Hybrid Automata [2], for which several approaches exist to analyze their time-bounded reachability, e.g., flowpipe construction for different state-space representations [9, 20, 23]. Several approaches for Hybrid Automata extended with discrete probability distributions exist [19, 28, 29, 31]. Stochastic hybrid systems require a high level of abstraction [1, 16], as e.g. applied to uncountable-state stochastic processes [27] and infinite-state Markov chains [17]. Related Petri net approaches [5, 12] are restricted e.g., with respect to the number of continuous variables [12] or to Markovian jumps [5].

Hybrid Petri nets with stochastic firings have shown to be useful for evaluating e.g. the *survivability* of critical infrastructures, like water and power distribution [6, 15] using model checking algorithms. Properties of HPnGs can be specified using Stochastic Time Logic (STL).

State space representation Choosing a state space representation that separates the stochastic from the deterministic evolution in the Petri net, their analysis has in the past been limited with respect to the number of stochastic firings. Since every firing of a general transition dimension adds one dimension to the state space, techniques and implementations for multi-dimensional geometric operations are required for the development of efficient and automated model checking techniques for HPnGs. While earlier work [7] was restricted to one stochastic firing, first extensions to more random variables [8, 10] required libraries for halfspace intersection and hyperplane arrangement, which to the best of our knowledge is not available for more than three dimensions.

The evolution of the state space over time of an HPnG can be partitioned into sets of states with similar behavior. This is done by conditioning their evolution on the firing times of the general transitions, either as *locations*, organized in a Parametric Location Tree (PLT) or using a geometric representation as convex polytopes (so-called regions) with similar characteristics. Recently vertex enumeration was proposed [14] to circumvent the problem of hyperplane arrangement, when constructing the geometric state set representation as regions.

The tree-based representation of the state space can be constructed for an arbitrary but finite number of stochastic firings [13]. In a next step, a geometric representation can be computed for each location in the PLT [14], featuring the library HyPro [26], which offers efficient implementations for operations on convex polytopes [32] in higher dimensions. The advantage of HyPro e.g. with respect to [4, 30], is the clean interface and the variety of options.

Model checking Stochastic Time Logic, which can be used to formally specify properties of the HPnG, closely resembles MITL [3] or the *temporal layer* of STL/PSL [22] and is used to specify properties of HPnGs. Model checking STL then relies on the geometric representation of regions as convex polytopes and recursively follows the parse tree of the formula. Per region a convex representation of its satisfying parts is returned. The satisfaction set of (the conjunction of) atomic properties is a single convex polytope, and *negation* requires a translation into a convex representation, since the representation of convex polytopes is not closed w.r.t. negation. Model checking the *until* operator relies on a series of geometric operations, including polytope inversion, and potentially iterates over child locations until the pre-specified time bound is reached. The resulting satisfaction sets implicitly contain the stochastic evolution and can be used to compute the probability that the STL property holds.

Integration Each random variable in the Hybrid Petri net is assigned a unique probability density function, which allows to integrate the joint probability distribution of all stochastic firings over the convex polytopes of the satisfaction set of an STL formula. Using the order in which the stochastic firings have occured, a d-dimensional Delayney Triangulation is required to create simplices of the convex polytopes from the satisfaction set. The ordered integration bounds are then obtained from an affine transformation of these simplices. To calculate the actual value of the integral, currently Monte Carlo methods [21] are used.

Tool support The analytical evaluation of HPnGs highly depends on their piece-wise linear evolution, resulting in exact state-space representations via convex polytopes. The recently released C++ tool hpngm efficiently implements and combines algorithms for state-space creation, transformation to a geometric representation, model checking a potentially nested STL formula and integrating over the resulting satisfaction set to yield the probability that the property holds. Since model checking and integrating are computationally expensive, the option of parallel execution has been included in the tool.

Furthermore, discrete-event simulation as proposed for Hybrid Petri nets [25] allows to validate analytical results and recent extensions to linear time invariant systems [18], enable the simulation of more complex continuous dynamics [24].

References

1. Abate, A., Katoen, J.-P., Lygeros, J., Prandini, M.: Approximate model checking of stochastic hybrid systems. Eur. J. Control **6,** 624–641 (2010)
2. Alla, H., David, R.: Continuous and hybrid petri nets. J. Circuits Syst. Comput. **8**(01), 159–188 (1998)
3. Alur, R., Feder, T., Henzinger, T.A.: The benefits of relaxing punctuality. J. ACM **43**(1), 116–146 (1996)
4. Bagnara, R., Hill, P.M., Zaffanella, E.: The parma polyhedra library: toward a complete set of numerical abstractions for the analysis and verification of hardware and software systems. Sci. Comput. Prog. **72**(1–2), 3–21 (2008)
5. Everdij, M.H.C., Blom, H.A.P.: Piecewise deterministic Markov processes represented by dynamically coloured Petri nets. Stochastics **77**(1), 1–29 (2005)
6. Ghasemieh, H., Remke, A., Haverkort, B.: Survivability analysis of a sewage treatment facility using hybrid petri nets. In: Performance Evaluation, vol. 97, pp. 36–56. Elsevier (2016)
7. Ghasemieh, H., Remke, A., Haverkort, B., Gribaudo, M.: Region-based analysis of hybrid petri nets with a single general one-shot transition. In: Jurdziński, M., Ničković, D. (eds.) FORMATS 2012. LNCS, vol. 7595, pp. 139–154. Springer, Heidelberg (2012). https://doi.org/10.1007/978-3-642-33365-1_11
8. Ghasemieh, H., Remke, A., Haverkort, B.R.: Hybrid petri nets with multiple stochastic transition firings. In: 2014 8th International Conference on Performance Evaluation Methodologies and Tools, pp. 217–224. ICST (2014)
9. Girard, A.: Reachability of uncertain linear systems using zonotopes. In: Morari M., Thiele L. (eds.) HSCC 2005. LNCS, vol. 3414, pp. 291–305. Springer, Heidelberg (2005). https://doi.org/10.1007/978-3-540-31954-2_19
10. Godde, A., Remke, A.: Model checking the STL time-bounded until on hybrid petri nets using net polyhedra. In: Reinecke, P., Di Marco, A. (eds.) EPEW 2017. LNCS, vol. 10497, pp. 101–116. Springer, Cham (2017). https://doi.org/10.1007/978-3-319-66583-2_7
11. Gribaudo, M., Remke, A.: Hybrid petri nets with general one-shot transitions. Perform. Eval. **105**, 22–50 (2016)
12. Horton, G., Kulkarni, V.G., Nicol, D.M., Trivedi, K.S.: Fluid stochastic Petri nets: Theory, applications, and solution techniques. J. Oper. Res. **105**(1), 184–201 (1998)
13. Hüls, J., Pilch, C., Schinke, P., Delicaris, J., Remke, A.: State-space construction of hybrid petri nets with multiple stochastic firings. In: Parker, D., Wolf, V. (eds.) QEST 2019. LNCS, vol. 11785, pp. 182–199. Springer, Cham (2019). https://doi.org/10.1007/978-3-030-30281-8_11

14. Hüls, J., Schupp, S., Remke, A., Ábrahám, E.: Analyzing hybrid petri nets with multiple stochastic firings using HyPro. In: 11th International Conference on Performance Evaluation Methodologies and Tools (2017)
15. Jongerden, M.R., Hüls, J., Remke, A., Haverkort, B.R.: Does your domestic photovoltaic energy system survive grid outages? Energies **9**(9), 736 (2016)
16. Julius, A.A.: Approximate abstraction of stochastic hybrid automata. In: Hespanha, J.P., Tiwari, A. (eds.) HSCC 2006. LNCS, vol. 3927, pp. 318–332. Springer, Heidelberg (2006). https://doi.org/10.1007/11730637_25
17. Klink, D., Remke, A., Haverkort, B.R., Katoen, J.-P.: Time-bounded reachability in tree-structured QBDs by abstraction. J. Spec. Issue Perform. Eval. **68**(2), 105–125 (2011)
18. Kofman, E.: A second-order approximation for DEVS simulation of continuous systems. SIMULATION **78**(2), 76–89, 2002.
19. Kwiatkowska, M., Norman, G., Segala, R., Sproston, J.: Automatic verification of real-time systems with discrete probability distributions. Theor. Comput. Sci. **282**(1), 101–150 (2002)
20. Le Guernic, C., Girard, A.: Reachability analysis of linear systems using support functions. Nonlinear Anal. Hybrid Syst. **4**(2), 250–262 (2010)
21. Peter Lepage, G.: A new algorithm for adaptive multidimensional integration. J. Comput. Phys. **27**(2), 192–203 (1978)
22. Maler, O., Nickovic, D.: Monitoring temporal properties of continuous signals. In: Lakhnech, Y., Yovine, S. (eds.) FTRTFT 2004, FORMATS 2004. LNCS, vol 3253, pp. 152–166. Springer, Heidelberg (2004). https://doi.org/10.1007/978-3-540-30206-3_12
23. Moore, R.E., Kearfott, R.B., Cloud, M.J.: Introduction to interval analysis. SIAM (2009)
24. Pilch, C., Niehage, M., Remke, A.: HPnGs go non-linear: statistical dependability evaluation of battery-powered systems. In: IEEE International Symposium on the Modeling, Analysis, and Simulation of Computer and Telecommunication Systems, pp. 157–169. IEEE (2018)
25. Pilch, C., Remke, A.: Statistical model checking for hybrid petri nets with multiple general transitions. In: 47th International Conference on Dependable Systems and Networks, pp. 475–486. IEEE (2017)
26. Schupp, S., Ábrahám, E., Makhlouf, I.B., Kowalewski, S.: HyPro: A C++ library of state set representations for hybrid systems reachability analysis. In: Barrett, C., Davies, M., Kahsai, T. (eds.) NFM 2017. LNCS, vol. 10227, pp. 288–294. Springer, Cham (2017). https://doi.org/10.1007/978-3-319-57288-8_20
27. Soudjani, S.E.Z., Gevaerts, C., Abate, A.: FAUST^2: formal abstractions of uncountable-STate STochastic processes. In: Baier, C., Tinelli, C. (eds.) TACAS 2015. LNCS, vol. 9035, pp. 272–286. Springer, Heidelberg. https://doi.org/10.1007/978-3-662-46681-0_23 (2015)
28. Sproston, J.: Decidable model checking of probabilistic hybrid automata. In: Joseph, M., (eds.) FTRTFT 2000. LNCS, vol. 1926, pp. 31–45. Springer, Heidelberg. https://doi.org/10.1007/3-540-45352-0_5 (2000)
29. Teige, T., Fränzle, M.: Constraint-based analysis of probabilistic hybrid systems. IFAC Proc. Vol. **42**(17), 162–167 (2009)
30. The CGAL Project. CGAL User and Reference Manual. CGAL Editorial Board, 4.10 edition (2017)
31. Zhang, L., She, Z., Ratschan, S., Hermanns, H., Hahn, E.M.: Safety verification for probabilistic hybrid systems. Eur. J. Control **18**(6), 572–587 (2012)
32. Ziegler, G.M.: Lectures on Polytopes, vol. 152. Springer Science & Business Media, New York (2012). https://doi.org/10.1007/978-1-4613-8431-1

Contents

Abandonment Attack on the LEACH Protocol 1
  Albatool Alhawas and Nigel Thomas

Coherent Resolutions of Nondeterminism 16
  Marco Bernardo

Emulating Self-adaptive Stochastic Petri Nets 33
  Lorenzo Capra and Matteo Camilli

Design and Evaluation of an Edge Concurrency Control Protocol
for Distributed Graph Databases 50
  Paul Ezhilchelvan, Isi Mitrani, Jack Waudby, and Jim Webber

A Novel Data-Driven Algorithm for the Automated Detection
of Unexpectedly High Traffic Flow in Uncongested Traffic States 65
  *Bo Klaasse, Rik Timmerman, Tessel van Ballegooijen, Marko Boon,
  and Gerard Eijkelenboom*

A Network Aware Resource Discovery Service 84
  Luigi Liquori, Rossano Gaeta, and Matteo Sereno

EthExplorer: A Tool for Forensic Analysis of the Ethereum Blockchain 100
  Yuriy Marchenko, William J. Knottenbelt, and Katinka Wolter

A Queueing Model that Works Only on the Biggest Jobs 118
  Andrea Marin and Sabina Rossi

Performance Evaluation of Thermal-Constrained Scheduling Strategies
in Multi-core Systems 133
  *Muhammad Usama Sardar, Clemens Dubslaff, Sascha Klüppelholz,
  Christel Baier, and Akash Kumar*

Bounding the Rate of Convergence for One Class of Finite Capacity
Time Varying Markov Queues 148
  *Alexander Zeifman, Yacov Satin, Rostislav Razumchik,
  Anastasia Kryukova, and Galina Shilova*

Author Index 161

Abandonment Attack on the LEACH Protocol

Albatool Alhawas[1,2(✉)] and Nigel Thomas[1]

[1] School of Computing, Newcastle University, Newcastle upon Tyne, UK
{a.alhawas2,nigel.thomas}@newcastle.ac.uk
[2] College of Computer and Information Sciences, King Saud University,
Riyadh, Kingdom of Saudi Arabia
Aalhawas2@ksu.edu.sa

Abstract. Despite their popularity and widespread use, wireless sensor networks are vulnerable to different types of attacks due to their low energy consumption, simplicity and scalability constraints. This paper explores the possible network-layer attacks on WSN routing protocols. In addition, it proposes a comprehensive method for measuring the impact of network-layer attacks on WSNs. Moreover, it introduces a new network-layer attack – called "abandonment attack" – on one such WSN routing protocol, the low-energy adaptive clustering hierarchy protocol. Last, it measures the impact of the abandonment attack on the LEACH protocol. In the end, this paper finds that the abandonment attack increases the collision rate and the end-to-end delay on the LEACH protocol and decreases the network lifetime.

Keywords: LEACH · Routing protocols · Security · Performance · Network attacks

1 Introduction

Wireless sensor networks (WSNs) grew in popularity after the emergence of internet of things (IoT) technology due to their unique properties of being cheap, simple, scalable and low in energy consumption. However, having such properties renders WSNs vulnerable to a wide range of attacks on all network layers: the physical layer, the data link layer, the network layer, the transport layer and the application layer. Therefore, attacks on WSNs should be examined and analysed to determine appropriate detection and prevention solutions. This study introduces a new network-layer attack, abandonment attack, on one specific WSN protocol: the low-energy adaptive clustering hierarchy (LEACH) protocol. The study examines the attack's impact on the LEACH protocol over various network settings (i.e., attack scenarios) to detect the influence of certain network properties, such as the malicious node location and the base station location, on the impact of the attack. This makes the study results more precise and nuanced.

The remainder of the paper is organised as follows: the background provides a brief overview of LEACH protocol and outlines current network-layer attacks on WSNs. The literature review surveys different related studies, proposes a comprehensive method for measuring the impact of network-layer attacks on WSNs and compares the current

© Springer Nature Switzerland AG 2020
M. Gribaudo et al. (Eds.): EPEW 2019, LNCS 12039, pp. 1–15, 2020.
https://doi.org/10.1007/978-3-030-44411-2_1

study with the literature. The abandonment attack section gives a theoretical explanation of abandonment attack. The simulation environment section provides details about the simulation settings and lists the attack scenarios. The results and evaluation section presents and analyses the impact of abandonment attack on the LEACH protocol. Finally, the paper ends with a conclusion and some suggestions for future work.

2 Background

2.1 The Low-Energy Adaptive Clustering Hierarchy Protocol

LEACH is a continuous protocol in which the sensor nodes periodically send their data to the base station – the base station does not query the sensor nodes for any data. It operates on rounds that keep running until all the sensor nodes on the network are dead. Each round is composed of a setup phase and a steady phase. In the setup phase, the sensor nodes form a collection of clusters – each one of which has a sensor node as its cluster head and multiple sensor nodes as child nodes. In the steady phase, which is composed of multiple time frames, the child nodes send their data to their correspondent cluster head. The cluster heads aggregate the data received from their child nodes and forward them to the base station at the end of every time frame. The steady phase keeps running until a new round begins with a new cluster setup. Having new cluster heads in each round deters the network from having dead cluster heads, allowing it to live longer.

The setup phase of the LEACH protocol is composed of the following steps: (1) determining cluster heads, (2) announcing cluster heads, (3) joining clusters, (4) creating a transmission schedule and (5) advertising the transmission schedule. In the first step, determining cluster heads, if all the nodes in the network have the same initial energy, a threshold value of $T(n, k, r)$ is calculated by all the nodes in the network according to Eq. 1. Here, n is the total number of nodes in the network, k is the maximum number of clusters in the network and r is the number of rounds that have passed. However, if the nodes have variant initial energies and a different threshold value, $T(E_i, E_{total}, k)$ is calculated by all the nodes in the network according to Eq. 2. Here, E_i is the energy of the node, E_{total} is the total energy for all the nodes in the network and k is the maximum number of clusters in the network.

$$T(n, k, r) = k/(n - k * (r \bmod n/k)) \tag{1}$$

$$T(E_i, E_{total}, k) = E_i * k/E_{total} \tag{2}$$

Next, all the nodes that have not been cluster heads for the past n/k rounds will choose a random value x from 0 to 1; if the node has been a cluster head, x will be 0. In the end, if x is less than the threshold, and the number of cluster heads in the network is less than k, the node will become a cluster head and will go to step two, announcing cluster heads, to announce itself as a cluster head by broadcasting an ADV message using a non-persistent carrier-sense multiple access (CSMA) MAC protocol. Here, CSMA means that the cluster heads sense the channel to check if there is another transmission before broadcasting the ADV message. If the node has not become a cluster head, it will go to step three, joining clusters, where it will listen for the ADV messages from the cluster heads and then send a join request message (Join-REQ) to the cluster head that has the strongest ADV signal (i.e., the closest cluster head) using a non-persistent CSMA MAC protocol. The node will then wait for a transmission schedule from its correspondent cluster head.

Once all the Join-REQ messages are sent, steps four and five will start, respectively. In other words, the cluster heads will create a transmission schedule and then advertise it by broadcasting an ADV-SCH message to their child nodes. This transmission schedule will eliminate internal collisions within the cluster by providing time division multiple access (TDMA) between the child nodes in the cluster. In addition to TDMA, a code division multiple access (CDMA) will be used by each cluster to avoid external collisions with other clusters in the network. Hence, when a cluster head and its child nodes exchange messages, they will spread their signals over a unique spreading code using a direct sequence spread spectrum (DSSS). Any messages that are spread over a different code will be considered as noise; thus, the message signals of one cluster will not collide with the message signals of another cluster.

After that, the setup phase ends and the steady phase begins. In the steady phase, the child nodes send their data in a DATA message to their correspondent cluster heads according to the transmission schedule of their cluster. The data will be sent using CDMA MAC protocol, which eliminates the collision rate and reduces the packet delay. Once a cluster head receives data from all its child nodes, it will aggregate the data into one message and transmit it to the base station using non-persistent CSMA MAC protocol. This transmission process will keep repeating itself until a new setup phase begins or until the end of the network lifetime.

In LEACH protocol, when the sensor nodes communicate with the base station or send an ADV message, they only use CSMA to sense whether the channel is idle before transmitting. Thus, collision might occur when two nodes sense the channel at the same time, and both find it idle. Also, the packet delay increases when there is a high number of nodes using the channel. Yet, when the sensor nodes communicate with their cluster heads, they use CDMA and TDMA, which eliminates the collision rate and reduces the packet delay [1].

2.2 Network Layer Attacks on WSNs

Due to their widespread use and design simplicity, WSNs have become a tempting target for attackers and, hence, a trending research topic for many security experts. In one study about security measurements in WSNs, Xie et al. [2] listed the five network layers of WSNs (the physical layer, the data link layer, the network layer, the transport layer and the application layer) and their associated threats and identified eight specific attacks against the network layer:

- **Replay attack.** The attacker catches a legitimate packet and re-sends it to different nodes in the network to consume their energy.
- **Sybil attack.** A malicious node possesses different identities and tricks a genuine node into falsely believing that it has multiple neighbours.
- **Blackhole attack.** A malicious node does not forward the packets it receives to their intended destinations.

- **Grayhole attack.** A malicious node does not forward some of the packets it receives to their intended destinations. This attack is also referred to as a 'selective forwarding attack'.
- **Wormhole attack.** Two malicious nodes in two remote locations form a tunnel between each other to forward legitimate packets to a different part of the network. This attack is considered to be particularly dangerous because it can forward legitimate packets without compromising any cryptography techniques.
- **Sinkhole attack.** A malicious node prompts the surrounding nodes to send it their packets by advertising a stronger signal and faster route. This attack is often used in conjunction with other types of attacks, such as blackhole and grayhole. Nevertheless, it differs from blackhole and grayhole attacks because in these attacks, the attacker does not change the route of the packets and may only discard the packets. However, in a sinkhole attack, the attacker may not discard the packets and may only increase the end-to-end delay.
- **Hello flood attack.** This attack targets routing protocols with hello messages (such as the LEACH protocol in Sect. 2.1, where cluster heads advertise themselves so that other nodes mark them as heads and forward their packets to them). In this attack, a malicious node will advertise itself as a cluster head, prompting other nodes to forward their packets to it.
- **Spoofing attack.** The attacker alters the routing information to increase end-to-end delays.

Table 1. Attacks against select routing protocols [6]

Protocol	Relevant attacks
TinyOS beaconing	Bogus routing information, selective forwarding, sinkhole, Sybil, wormhole, hello flood
Directed diffusion and its multipath variant	Bogus routing information, selective forwarding, sinkhole, Sybil, wormhole, hello flood
Geographic routing (GPSR, GEAR)	Bogus routing information, selective forwarding, Sybil
Clustering-based protocols (LEACH, TEEN, PEGASIS)	Selective forwarding, hello flood
Rumour routing	Bogus routing information, selective forwarding, sinkhole, Sybil, wormhole
Energy-conserving topology maintenance (SPAN, GAF, CEC, AFEA)	Bogus routing information, Sybil, hello flood

Alajmi, Pathan et al. and Goyal et al. described similar possible network layer attacks in [3, 4] and [5], respectively. Goyal et al. [5] also included acknowledgment of spoofing as a possible attack. In this method, the attacker sends a replayed or forged acknowledgment to deceive a genuine node into believing that a dead node is alive or that a weak link is strong.

Karlof and Wagner [6] evaluated the possibility of different attacks against multiple routing protocols. Their study mainly focused on Berkeley's TinyOS sensor platform. Table 1 summarises the list of attacks they evaluated against each protocol ('bogus routing information' refers to spoofing and replay attacks, and 'selective forwarding' includes both blackhole and grayhole attacks). In Table 1, one notices that they claimed that clustering-based protocols, such as LEACH, were only subjected to selective forwarding and hello flood attacks. Yet according to the LEACH description in Sect. 2.1, it appears that compromised cluster heads could perform Sybil and wormhole attacks. This is perhaps because the authors assumed that all the cluster heads were benign.

3 Literature Review

It is important to consider the studies that examine WSN performance while under attack to find appropriate ways to analyse such attacks. The first such study is by Almomani et al. [7]; it examined LEACH protocol performance under hello flood, grayhole, blackhole and scheduling attacks. Scheduling attacks are a new type of attack that the aforementioned researchers introduced in their paper. Here, a malicious cluster head in the LEACH protocol amends the packet transmission time of its child nodes so that all its child nodes send their packets at the same time, causing an intentional collision. In their study, the researchers simulated the attacks using an NS-2 simulator on one fixed network topology composed of 100 nodes and five cluster heads. The simulation of the attacks was performed with three different levels of compromised nodes in the network: 10%, 30% and 50%. After the simulation, the effects of the attacks on the network were presented using three matrices (packet delivery ratio, network lifetime and consumed energy). Overall, they found that the flood attack decreased the network lifetime and the packet delivery ratio. The blackhole, grayhole and scheduling attacks increased the network lifetime and decreased the packet delivery ratio.

Ioannou et al. [8][1] measured the impact of grayhole, grayhole plus sinkhole and hello flood attacks on a weighted shortest path (WSP), a protocol that propagates messages between nodes through a path that is built based on the nodes' distances from the final destination and the nodes' signal strength, respectively. In this study, they used a COOJA simulator to model the attacks on two network topologies comprised of 25 nodes each. One topology located the base station in the middle of the network, whereas the other topology had the base station at the edge of the network. All the simulation scenarios contained only one malicious node. However, they simulated the attack 25 times (the number of nodes in the network), with each simulation having a different malicious node location. The researchers found that in grayhole attacks, packet loss was highly significant if the malicious node was next to the base station. However, its significance

---

[1] The researchers used different names for the attacks in their study.

decreased dramatically when the malicious node was further away from the base station. They also noted that the network topology with the base station at the edge was more affected by the grayhole attack than the network topology with the base station in the middle. In addition, they found that a combined grayhole and sinkhole attack always had significant packet loss regardless of the malicious node location or the base station location (because in a sinkhole attack, a malicious node deceives other nodes into sending it their packets). In the hello flood attack, the results were arbitrary for the different malicious node locations.

Salam et al. [9] used an NS-2 simulator to analyse the impact of a hello flood attack on an ad hoc on-demand distance vector (AODV) protocol. This protocol is similar to directed diffusion protocol, a well-known WSN protocol, but it allows a source node to build a route to a destination node only when necessary and does not maintain routes that are not currently in use. In the study, a hello flood attack was simulated on one fixed network topology of 100 nodes with one, four, five and six malicious nodes. The network throughput and packet delay matrices were used to measure the impact of the hello flood attack on the WSNs. The results showed that as the number of malicious nodes increased, the packet delay increased and the throughput decreased.

Lastly, Baskar et al. [10] used an NS-2 simulator to simulate a sinkhole attack on a tree-based routing protocol that is described in their paper. The researchers used energy consumption, throughput and a packet delivery ratio as matrices to measure the impact of sinkhole attacks in different attack scenarios that varied in network size, number of malicious nodes, location of malicious nodes and power of malicious nodes. The results showed that the impact of a sinkhole attack did not change with the change in the network size; however, it increased with the increase in the number of malicious nodes or with the increase of the power of the malicious nodes. Also, the impact of the sinkhole attack increased when the malicious node was closer to the base station.

Table 2 summarises the aforementioned studies. The table shows the protocols, attacks, impact matrices and number of topologies used in the studies. In addition, it indicates whether the studies covered different amounts of malicious nodes, different malicious node locations and different malicious node powers. Based on this data, four parameters impacted the results of the cited studies and must be taken into consideration when an attack is being simulated: (1) the network topology (i.e., location of the base station), (2) the number of malicious nodes, (3) the location of the malicious nodes and (4) the power of the malicious nodes. Also, there are several ways to measure the impact of an attack on WSNs. These include packet delivery ratio (PDR), energy consumption (EC), network lifetime (LT), throughput and packet delay.

It is, however, advisable to look at all these matrices when an attack impact analysis is being conducted, as looking at the results of [10] and [8], there is a contrast in their results that could be due to the different impact matrices used in the studies. In [10], the researchers stated that the significance of the sinkhole attack increased when the malicious node moved closer to the base station, whereas in [8], the researchers stated that the sinkhole attack had a significant impact on the network regardless of the location of the malicious node. This contrast could be because in [8], they only considered the packet delivery ratio, whereas in [10], they considered the energy consumption, throughput and packet delivery ratio.

Table 2. Literature review with comparison of the present study

Ref.	Protocol	Attacks	Impact matrices	Network topology	Different number of malicious nodes	Different location of malicious nodes	Different power of malicious nodes
[7]	LEACH	Blackhole, grayhole, hello flood, scheduling	PDR, EC, LT	One	Yes	No	No
[8]	WSP	Grayhole, sinkhole plus grayhole, hello flood	PDR	Two	No	Yes	No
[9]	AODV	Hello flood	Throughput and delay	One	Yes	No	No
[10]	TBR	Sinkhole	PDR, EC, throughput	One with two different sizes	Yes	Yes	Yes
This paper	LEACH	Abandonment	PDR, throughput, delay, LT, EC	Two	Yes	No	Yes

Table 2 also shows the difference between the present study and the literature. This study introduces the abandonment attack – a new network layer attack on the LEACH protocol – and measures its impact on the LEACH protocol's PDR, EC, network LT, throughput and end-to-end delay. In addition, this study uses different attack scenarios to show the influence of the network topology, number of malicious nodes and power of the malicious nodes on the analysis results.

4 Abandonment Attack

This study introduces a new network layer attack on the LEACH protocol: abandonment attack. In this attack, a malicious cluster head sends an ADV message to announce that it has become a cluster head. It then receives Join-REQ messages from child nodes. Instead of creating a legitimate transmission schedule, it creates a fake transmission schedule full of fake node IDs to deceive its child nodes into thinking that they do not have a transmission time. This being done, the child nodes transmit their data directly to the base station using a non-persistent CSMA MAC protocol, which causes them to consume more energy because the base station is far from these nodes, and sending data to a far location is energy consuming. It also increases both the collision rate and the transmission delay since the number of nodes directly connected to the base station increases. That is because all these nodes that are directly connected to the base station transmit their signals using CSMA, in which the nodes keep sensing the channel until it is idle to transmit their signals – if two nodes transmit their signals at the same time, a collision occurs. This is unlike when the child nodes are connected to a cluster head, where they use TDMA and CDMA to avoid inter-cluster and intra-cluster collisions and to minimise the transmission delay.

5 Simulation Environment

The attack was simulated using an NS-2.34 simulator on top of Wendi Heinzelman's code of LEACH protocol [11]. The attack was run over different simulation scenarios to capture the influence of some network properties (i.e. base station location, number of malicious nodes and power of malicious nodes) on the impact of the attacks. This section explains these different scenarios in detail. Also, Table 3 shows the general proprieties that are shared across all the scenarios.

Table 3. General simulation properties

Network properties	
Network size	100 m × 100 m
Number of nodes	100 nodes
Maximum number of clusters[a]	5 clusters
Data size[b]	500 bytes
Header size	25 bytes
Network bandwidth	1 Mbps
Round duration (cluster change)	20 s
Simulation time	500 s
Sensor properties	
Sensor type	μAMPS
Idle energy	0
Sleep energy	0
Beam-forming energy	5e-9 J/Bit
Transmit amplifier energy	9.67e-12 J/Bit/m^2

[a]This is the optimum number of cluster heads to get the best energy dissipation results in the current network settings [12].
[b]NS-2 does not send real packets; therefore, the data size and the header size are fixed.

5.1 Scenario I: Different Malicious Node Percentages

In this scenario, the LEACH performances under attack with 0%, 10%, 20% and 30% malicious nodes in the network are compared. The base station is located at the top of the network at point (50, 175), and all the sensor nodes have an initial energy of two joules. Figure 1 shows the locations of the malicious nodes in the network.

5.2 Scenario II: Different Base Station Locations

This scenario compares the LEACH performance under attack when the base station is located in the middle of the network at point (50, 50) with the LEACH performance under attack when the base station is located at the top of the network at point (50, 175). All the sensor nodes in this scenario have an initial energy of two joules, and the percentage of malicious nodes in the network is 10%. Figures 1 and 2 show the locations of the base station and the malicious nodes.

5.3 Scenario III: Malicious Nodes Have Higher Energy

This scenario compares the LEACH performance under attack when the malicious nodes have an initial power that is equal to the genuine sensor nodes, which is two joules of energy, with the LEACH performance under attack when the malicious nodes have a higher power than the genuine sensor nodes, which is 200 J of energy. Based on the LEACH protocol explanation in Sect. 2.1, the cluster head selection process will be different when the nodes in the network have different initial powers. Thus, the nodes with a higher power will have a higher probability of becoming a cluster head. In any case, the location of the base station in this scenario is at the top of the network at point (50, 175), and the percentage of malicious nodes in the network is 10%. Figures 1 and 2 show the locations of the base station and the malicious nodes.

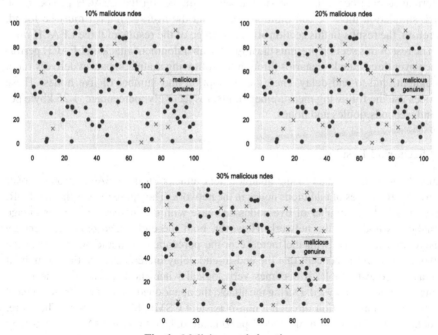

Fig. 1. Malicious node locations

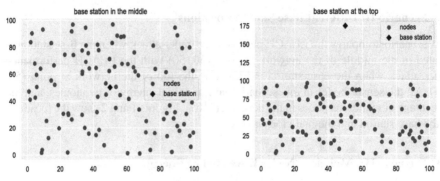

Fig. 2. Base station locations

6   Results and Evaluation

In the LEACH protocol, the effect of an attack cannot be measured by only looking at the results of a single run, for two reasons. First, the cluster head locations and the number of child nodes in each cluster have a major effect on the performance matrices in the LEACH protocol. Second, the cluster head locations and the number of child nodes in each cluster change with every round in a single LEACH run. Hence, each run of the LEACH protocol will have different cluster head combinations and will yield different performance matrices. So, to measure an attack impact on the LEACH protocol, the attack simulation must be run multiple times and the average impact must be found. Therefore, the results in this section are the average of the results of three LEACH runs.

The rest of this section presents the impact of abandonment attack on LEACH performance over three attack scenarios through the following matrices: packet delivery ratio, throughput, end-to-end delay, energy consumption and number of live nodes. These matrices use the following measurement units, respectively: percentage, 1.25 kbyte per second, seconds, joules and nodes.

6.1   Scenario I Results

Figure 3 shows the impact of the abandonment attacks on the LEACH protocol over different percentages of malicious nodes in the network. The figure reveals that the PDR, throughput and the number of live nodes decrease with the increase in the percentage of malicious nodes in the network. However, both the end-to-end delay and energy consumption level rise with the increase in the percentage of malicious nodes in the network. Though the concept that the end-to-end delay increases while the throughput decreases is questionable, it becomes very logical when the behaviour of the nodes during the attack is analysed. During the attack, the abandoned child nodes try to connect directly to the base station through a non-persistent CSMA MAC protocol. Therefore, the delay increases because the nodes' packets are delayed on the MAC layer, as they must wait for the channel to be clear before they are sent. In contrast, the throughput, which is measured by the number of delivered bits per second, decreases because only one packet can keep the whole channel busy. Without the abandonment attack, the nodes join different clusters and use CDMA and TDMA to send multiple packets at the same time with minimal delay. In addition, Fig. 3 illustrates that the energy consumption level increases with the increase in the percentage of malicious nodes in the network. This is because the increase in the number of the malicious nodes increases the possibility that a malicious node will become a cluster head. Hence, more nodes will be abandoned and will connect directly to the base station. Unlike the nodes that are connected to a cluster head, the nodes that are connected to the base station will always be awake and never go to sleep, causing them to consume more energy.

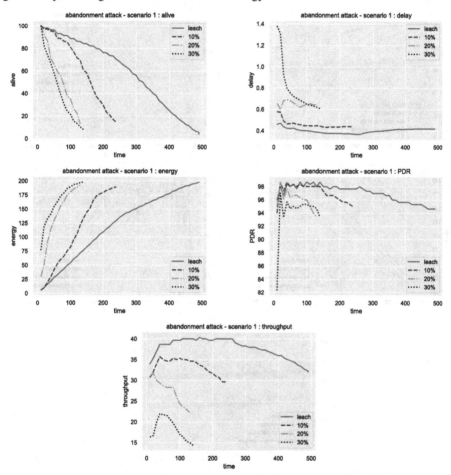

Fig. 3. Abandonment attack results – Scenario I

6.2 Scenario II Results

Figure 4 displays the impacts of the abandonment attacks when the base station is in the middle of the network and when the base station is at the top of the network. When the base station is at the top of the network, the abandonment attack has lower impact on the end-to-end delay. This is because when the base station is in the middle, and many abandoned nodes are sending their packets directly to the base station using non-persistent CSMA MAC protocol, the channel is busy most of the time and subjected to a lot of noise. However, when the base station is located at the top, the channel has less noise, causing the impact on the delay to be lower. Further, there is fluctuation in the PDR of the LEACH when the base station is in the middle. This fluctuation is because the performance parameters are logged every 10 s, with a new cluster head nomination process beginning every 20 s.

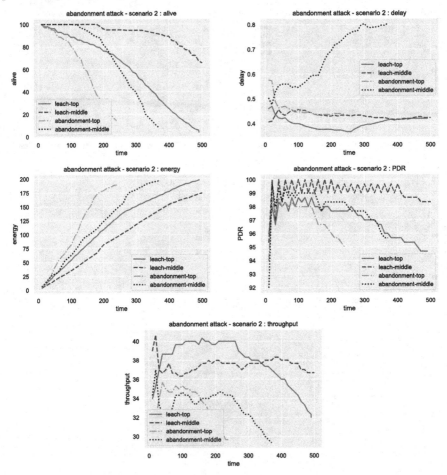

Fig. 4. Abandonment attack results – Scenario II

6.3 Scenario III Results

The LEACH protocol has two variations. In the first one, the nodes' energy is not a parameter in the cluster head nomination process. In the second one, the nodes' energy is a parameter in the cluster head nomination process. Figure 5 exhibits the impact of the abandonment attack on the second variation of the LEACH protocol and compares it to the impact of the attack on the first variation. Here, the 'power abandonment attack' refers to the second variation of LEACH, with the malicious nodes having a higher energy than the genuine nodes. The 'abandonment attack' refers to the first variation of LEACH, with the malicious nodes having an energy equal to the genuine nodes. In the power abandonment attack, only malicious nodes can be cluster heads because they have more energy than the genuine nodes. Therefore, in every round, all the child nodes are abandoned, leading to low PDR and throughput at all times. Moreover, in the power abandonment attack, the energy consumption level rises rapidly while the number of live nodes declines sharply. However, the number of live nodes stops decreasing when it reaches 10 nodes, as this is the number of malicious nodes in the network. These nodes have higher energy than all the other nodes, therefore they do not die quickly. They are, also, the reason why the network has a long lifetime and the end-to-end delay never drops below 0.8 s. Still, the end-to-end delay decreases gradually with the decrease in the number of live nodes in the network.

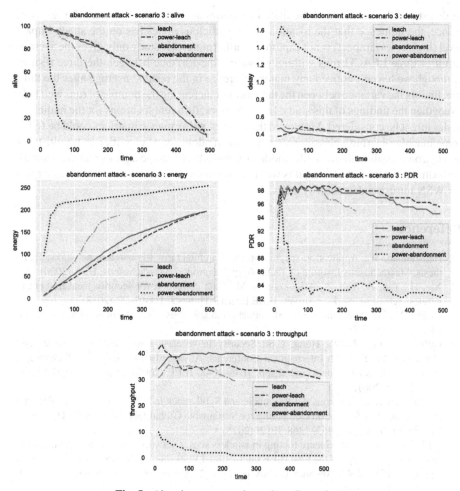

Fig. 5. Abandonment attack results – Scenario III

7 Conclusion

The study results elicit several points. First, the abandonment attack increases the collision rate, end-to-end delay, and the energy consumption level. Also, it decreases the PDR, throughput, and number of live nodes in the network. Second, in spite of the abandonment attack, the LEACH protocol maintains its property of having better performance when the base station is in the middle of the network. Yet, the impact of the abandonment attack on the end-to-end delay is higher when the base station is in the middle of the network. Third, in an abandonment attack, when the malicious nodes have higher energy then the genuine nodes in the network, they have a higher probability of becoming cluster heads than the genuine nodes, leading the impact of the attacks to be higher.

8 Limitations and Future Work

In this study, the locations of the malicious nodes are static in all the attack scenarios. Hence, the influence that the location of the malicious nodes has on the attack impact level (high or low impact) is not captured. Furthermore, this study only gives the average result of three LEACH runs without providing the variance between the three results. Although we haven't noticed any major difference in the results' average after the third run, finding the variance between the three runs or increasing the number of runs well help strengthen the findings of this study. In future work, different locations for the malicious nodes might be considered. Moreover, the attacks' performance matrices could be fed to some machine-learning classifiers, such as the random forest classifier and multi-layer perceptron classifier, to derive the hidden relations between the attacks and the network performance anomalies. Doing this will help researchers to create attack detection rules for WSN intrusion detection systems.

References

1. Heinzelman, W.: Application-specific protocol architectures for wireless networks. Ph.D. thesis, Massachusetts Institute of Technology, Cambridge, Massachusetts, USA (2000)
2. Xie, H., Yan, Z., Yao, Z., Atiquzzaman, M.: Data collection for security measurement in wireless sensor networks: a survey. IEEE Internet Things J. 1(1), 1–22 (2018)
3. Alajmi, N.: Wireless sensor networks attacks and solutions. Int. J. Comput. Sci. Inf. Secur. 12(7), 37–40 (2014)
4. Pathan, A., Lee, H.-W., Hong, C.S.: Security in wireless sensor networks: issues and challenges. In: 2006 8th International Conference Advanced Communication Technology, Phoenix Park, South Korea, vol. 2, pp. 1046–1048. IEEE (2006). https://doi.org/10.1109/ICACT.2006.206151
5. Goyal, S., Bhatia T., Verma, A.: Wormhole and Sybil attack in WSN: a review. In: 2015 2nd International Conference on Computing for Sustainable Global Development (INDIACom), New Delhi, India, pp. 1463–1468. IEEE (2015)
6. Karlof, C., Wagner, D.: Secure routing in wireless sensor networks: attacks and countermeasures. In: Proceedings of the First IEEE International Workshop on Sensor Network Protocols and Applications 2003, Anchorage, USA, pp. 113–127. IEEE (2003). https://doi.org/10.1109/SNPA.2003.1203362

7. Almomani, I., Al-Kasasbeh, B.: Performance analysis of LEACH protocol under Denial of service attacks. In: 2015 6th International Conference on Information and Communication Systems (ICICS), Amman, Jordan, pp. 292–297. IEEE (2015)

8. Ioannou, C., Vassiliou, V.: The impact of network layer attacks in wireless sensor networks. In: 2016 International Workshop on Secure Internet of Things (SIoT), Heraklion, Greece, pp. 20–28. IEEE (2016). https://doi.org/10.1109/SIoT.2016.009

9. Salam, M., Halemani, N.: Performance evaluation of wireless sensor network under hello flood attack. Int. J. Comput. Netw. Commun. (IJCNC) **8**(2), 78–87 (2016)

10. Baskar, R., Raja, K., Joseph, C., Reji, M.: Sinkhole attack in wireless sensor networks—performance analysis and detection methods. Indian J. Sci. Technol. **10**(12), 1–8 (2017)

11. μAMPS ns Code Extensions. http://www.mtl.mit.edu/researchgroups/icsystems/uamps/research/leach/leach_code.shtml. Accessed 06 May 2019

12. Heinzelman, W.R., Chandrakasan, A., Balakrishnan, H.: Energy-efficient communication protocol for wireless microsensor networks. In: Proceedings of the 33rd Annual Hawaii International Conference on System Sciences, Maui, HI, USA, pp. 1–10. IEEE (2000). https://doi.org/10.1109/HICSS.2000.926982

Coherent Resolutions of Nondeterminism

Marco Bernardo[✉]

Dipartimento di Scienze Pure e Applicate, Università di Urbino, Urbino, Italy
marco.bernardo@uniurb.it

Abstract. We study the impact that different ways of resolving nondeterminism within probabilistic automata have on the properties of probabilistic behavioral equivalences. Firstly, we provide a uniform definition of structure-preserving and structure-modifying resolutions of nondeterminism, respectively generated by different families of schedulers. Secondly, we exhibit a number of anomalies arising from the excessive power of the various families of schedulers, which affect the discriminating power, the compositionality, and the backward compatibility of probabilistic trace equivalence. Thirdly, we propose to remove those anomalies by enforcing coherency within resolutions of nondeterminism. This ensures that a scheduler cannot select different continuations in equivalent states of an automaton, so that also the states to which they correspond in any resolution of the automaton have equivalent continuations.

Keywords: Probabilistic automata · Schedulers · Equivalences

1 Introduction

Quantitative models of computing systems describe the order in which activities are executed – possibly admitting nondeterminism in case of concurrency phenomena or to support implementation freedom – and include information about the probabilities or the timing of the activities themselves. A particularly expressive model is given by *probabilistic automata* [22], as they encompass fully nondeterministic models like labeled transition systems [18], fully probabilistic models like action-labeled variants of discrete-time Markov chains [19], and reactive probabilistic models like Markov decision processes [11].

Behavioral relations play a fundamental role in the analysis of quantitative models. They formalize observational mechanisms that permit relating models that, despite their different representations in the same mathematical domain, cannot be distinguished by external entities when abstracting from details deemed unimportant for specific purposes. Moreover, they support system modeling and verification by providing a means to relate system descriptions expressed at different levels of abstraction, as well as to reduce the size of a system representation while preserving specific properties to be assessed later.

In the case of fully nondeterministic models, from the first comparative work [8] to the elaboration of the full spectrum [13], a number of equivalences

© Springer Nature Switzerland AG 2020
M. Gribaudo et al. (Eds.): EPEW 2019, LNCS 12039, pp. 16–32, 2020.
https://doi.org/10.1007/978-3-030-44411-2_2

have emerged that range from the branching-time – i.e., (bi)simulation-based – endpoint [21] to the linear-time – i.e., trace-based – endpoint [7] passing through testing relations [9]. The spectrum becomes simpler when considering fully probabilistic models [1,14,17], whereas as shown in [4] it is much more variegated in the case of models with nondeterminism and probabilities like probabilistic automata. The reason is that the probability of equivalence-specific events can be calculated only after removing nondeterminism. Examples of such events are reaching via given actions certain sets of equivalent states (bisimulation semantics) or executing specific action sequences (trace semantics), with states/traces being possibly decorated with additional information.

In this paper, we study the impact on the discriminating power, the compositionality, and the backward compatibility of behavioral equivalences for nondeterministic and probabilistic models, due to the different ways of resolving nondeterminism. We restrict ourselves to *simple* probabilistic automata [22], i.e., state-transition graphs where each transition is labeled with an action and goes from a state to a probability distribution over states. In this model, nondeterminism is expressed by the presence of *several* transitions departing from the same state. A *resolution of nondeterminism* is obtained by applying a *scheduler* that decides which activity has to be performed next, where by activity we mean executing a transition or stopping the execution altogether.

The first contribution of this paper is a discussion of different families of schedulers, with the result of providing a uniform way, based on *correspondence functions*, of defining the resolutions induced by those schedulers.

We divide resolutions into *structure preserving* and *structure modifying*, depending on whether they respect or alter the structure of the automaton from which they are obtained. A structure-preserving resolution is produced by a *deterministic scheduler*, which selects at the current state one of the transitions departing from that state or no transitions at all. A structure-modifying resolution is derived via a *randomized scheduler* [22], which probabilistically combines the transitions departing from the current state, or an *interpolating scheduler* [10], which splits the current state into copies, each having at most one outgoing transition, whose probabilities sum up to the probability of the original state. We formalize any resolution as a fully probabilistic automaton, which we equip with a correspondence function from the acyclic state space of the resolution to the possibly cyclic state space of the original automaton, as done for the first time in [15] for deterministic schedulers.

The second contribution of this paper is the presentation of a number of anomalies affecting probabilistic behavioral equivalences, mostly arising under deterministic schedulers, together with a proposal for avoiding them based on limiting the excessive power of schedulers.

We focus on probabilistic trace equivalence by showing that it does not contain probabilistic bisimilarity, it is not a congruence with respect to action prefix, and it is not backward compatible with its version for fully probabilistic models. The reason is that schedulers have the freedom to make *different* decisions in *equivalent* states occurring in the target distribution of a transition, with these decisions being not necessarily replicable in equivalent distributions of distinct automata. This is especially true for deterministic schedulers, as the resolutions they induce must be structure preserving.

Such anomalies can be avoided by employing *coherent resolutions* in the definition of probabilistic trace equivalence. The idea is that, if several states in the target distribution of a transition are equivalent, then the states to which they correspond in a resolution must be equivalent as well. This constraint can be formalized by reasoning on trace distributions, i.e., families of sets of traces each endowed with its execution probability in a given resolution.

This paper is organized as follows. In Sect. 2, we recall simple probabilistic automata. In Sect. 3, we discuss different notions of resolution usable in probabilistic behavioral equivalences and provide a uniform way of defining all of them. In Sect. 4, we illustrate the aforementioned anomalies of probabilistic trace equivalence caused by the excessive power of schedulers. In Sect. 5, we show how to avoid those anomalies by forcing resolutions to be coherent. Finally, in Sect. 6 we present some concluding remarks.

2 Nondeterministic and Probabilistic Models

We formalize systems featuring nondeterminism and probabilities through a variant of simple probabilistic automata [22], in which we do not distinguish between external and internal actions.

Definition 1. *A* nondeterministic and probabilistic labeled transition system, *NPLTS for short, is a triple* $(S, A, \longrightarrow)$ *where* $S \neq \emptyset$ *is an at most countable set of states,* $A \neq \emptyset$ *is a countable set of transition-labeling actions, and* $\longrightarrow \subseteq S \times A \times Distr(S)$ *is a transition relation with* $Distr(S)$ *being the set of discrete probability distributions over* S. ∎

A transition (s, a, Δ) is written $s \xrightarrow{a} \Delta$. We say that $s' \in S$ is not reachable from s via that a-transition if $\Delta(s') = 0$, otherwise we say that it is reachable with probability $p = \Delta(s')$. The reachable states form the support of Δ, i.e., $supp(\Delta) = \{s' \in S \mid \Delta(s') > 0\}$. An NPLTS can be depicted as a directed graph in which vertices represent states and action-labeled edges represent transitions, with states in the same support being linked by a dashed line and decorated with the respective probabilities (see the forthcoming Figs. 1, 2, 3, 4, 5, 6, 7, 8 and 9).

An NPLTS represents (i) a *fully nondeterministic* process when every transition has a target distribution with a singleton support, (ii) a *fully probabilistic* process when every state has at most one outgoing transition, or (iii) a Markov decision process when for each action any state has at most one outgoing transition labeled with that action implying the absence of *internal nondeterminism*.

Definition 2. *Let* $\mathcal{L} = (S, A, \longrightarrow)$ *be an NPLTS and* $s, s' \in S$. *We say that the finite sequence:*

$$c \equiv s_0 \xrightarrow{a_1} s_1 \xrightarrow{a_2} s_2 \dots s_{n-1} \xrightarrow{a_n} s_n$$

is a computation *of* $\mathcal{L}$ *of length* $n \in \mathbb{N}$ *from* $s = s_0$ *to* $s' = s_n$ *compatible with trace* $\alpha = a_1 a_2 \ldots a_n \in A^*$, *written* $c \in \mathcal{CC}(s, \alpha)$, *iff for all* $i = 1, \ldots, n$ *there exists in* $\mathcal{L}$ *a transition* $s_{i-1} \xrightarrow{a_i} \Delta_i$ *such that* $s_i \in supp(\Delta_i)$, *with:*

- $\Delta_i(s_i)$ *being the execution probability of step* $s_{i-1} \xrightarrow{a_i} s_i$ *conditioned on the selection of transition* $s_{i-1} \xrightarrow{a_i} \Delta_i$ *at state* s_{i-1}, *or simply the execution probability of that step if* $\mathcal{L}$ *is fully probabilistic;*
- $prob(c) = \prod_{1 \le i \le n} \Delta_i(s_i)$ *being the execution probability of* c *if* $\mathcal{L}$ *is fully probabilistic, assuming that* $prob(c) = 1$ *when* $n = 0$;
- $prob(C) = \sum_{c \in C} prob(c)$ *if* $\mathcal{L}$ *is fully probabilistic, provided that none of the computations in* C *is a proper prefix of one of the others.* ■

3   An Overview of Resolutions of Nondeterminism

When several transitions depart from the same state s of an NPLTS $\mathcal{L}$, they describe a nondeterministic choice among different behaviors. Eliminating these choices is necessary to perform the calculations required by probabilistic behavioral equivalences. A *resolution* of s is the result of a possible way of resolving nondeterministic choices starting from s, as if a *scheduler* were applied that decides which activity has to be performed next. A resolution of nondeterminism can thus be formalized as a *fully probabilistic* NPLTS $\mathcal{Z}$ with a *tree-like structure*, whose branching points correspond to target distributions of transitions deriving from those of $\mathcal{L}$.

We now present an overview of various ways of resolving nondeterminism, with the result of providing a uniform technique for defining all of them based on correspondence functions, so to facilitate their comparison. In Sects. 3.1 to 3.3 we address the notions of resolution stemming from two different approaches, respectively preserving or modifying the structure of the original NPLTS. The idea underlying the former approach is to construct a resolution by *importing states and transitions* from the original model. The idea at the basis of the latter approach is that (i) a transition of a resolution can be produced by *probabilistically combining transitions* of the original model, or (ii) a state of a resolution can be obtained by *probabilistically splitting states* of the original model.

3.1   Structure-Preserving Resolutions via Deterministic Schedulers

A *deterministic scheduler* selects one of the transitions departing from the current state or no transitions at all thus stopping the execution. As a consequence, the resulting resolution is isomorphic to a submodel of the original model (or of its unfolding, should cycles be present), thereby *preserving* the structure of the original model (or of its unfolding). If the model is fully nondeterministic, each of its resolutions coincides with a computation of the model; if the model is fully probabilistic, its maximal resolution coincides with the entire model.

In [26] a resolution was defined as a maximal subtree of the unfolding of the considered model – with the unfolding yielding a potentially infinite tree – in which every state has at most one outgoing transition. Resolutions were defined as fully probabilistic maximal subtrees also in [16], but the considered models were finite trees in lieu of directed graphs. Subtree maximality was required just because of the focus of those works on testing semantics.

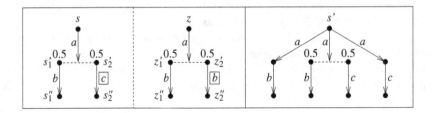

Fig. 1. Lack of injectivity breaks structure preservation

The paper [15], instead of reasoning in terms of unfoldings and submodels, introduced for the first time a *correspondence function* $corr_Z : Z \to S$ from the acyclic state space of the resolution $\mathcal{Z} = (Z, A, \longrightarrow_Z)$ being built, to the possibly cyclic state space of the considered model $\mathcal{L} = (S, A, \longrightarrow)$. This function had to satisfy the following constraint on transitions: if $z \xrightarrow{a}_Z \Delta$ then $corr_Z(z) \xrightarrow{a} \Gamma$, with $\Delta(z') = \Gamma(corr_Z(z'))$ for all $z' \in supp(\Delta)$.

The correspondence function with its constraint as defined in [15] and reused in [3,4] has the drawback of not being structure preserving in the case that the target distribution of a transition assigns the same probability to several inequivalent states. Let us see for instance the three NPLTS models in Fig. 1. The correspondence function that maps z to s, z'_1 and z'_2 to s'_1, and z''_1 and z''_2 to s''_1 causes the central NPLTS to be considered a legal resolution of the leftmost NPLTS, although the former is not isomorphic to any submodel of the latter. This may have no consequences on the discriminating power of testing equivalences, the subject of [15], if all transitions of testing systems are identically labeled. However, it would lead to consider the leftmost NPLTS and the rightmost NPLTS as trace equivalent, because also the leftmost one would have a resolution in which trace $a\,b$ (resp. trace $a\,c$) is executable with probability 1.

The constraint was rectified in [5] by requiring the *injectivity* of $corr_Z$ over $supp(\Delta)$, so that in Fig. 1 z'_1 and z'_2 can no longer be both mapped to s'_1. We also point out that in [2] it was further observed that *bijectivity* between $supp(\Delta)$ and $supp(\Gamma)$, rather than injectivity, is necessary to preserve the overall reachability mass in more general settings like the ULTRAS metamodel where, unlike the probabilistic case, there is no predefined value like 1 for the reachability mass of the target of a transition.

Below is the rectified definition of [5] in the style of [15], i.e., based on a correspondence function from the acyclic state space of the resolution to the possibly cyclic state space of the considered model.

Definition 3. *Let* $\mathcal{L} = (S, A, \longrightarrow)$ *be an NPLTS and* $s \in S$. *An acyclic NPLTS* $\mathcal{Z} = (Z, A, \longrightarrow_{\mathcal{Z}})$ *is a structure-preserving resolution of* s, *written* $\mathcal{Z} \in Res_{\mathrm{sp}}(s)$, *iff there exists a correspondence function* $corr_{\mathcal{Z}} : Z \to S$ *such that* $s = corr_{\mathcal{Z}}(z_s)$, *for some* $z_s \in Z$, *and for all* $z \in Z$ *it holds that:*

- *If* $z \stackrel{a}{\longrightarrow}_{\mathcal{Z}} \Delta$ *then* $corr_{\mathcal{Z}}(z) \stackrel{a}{\longrightarrow} \Gamma$, *with* $corr_{\mathcal{Z}}$ *being injective over* $supp(\Delta)$ *and satisfying* $\Delta(z') = \Gamma(corr_{\mathcal{Z}}(z'))$ *for all* $z' \in supp(\Delta)$.
- *At most one transition departs from* z. ∎

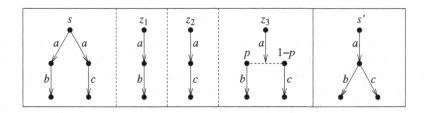

Fig. 2. An example of structure modification induced by a randomized scheduler

3.2   Structure-Modifying Resolutions via Randomization

If the current state has $n \in \mathbb{N}_{\geq 1}$ outgoing transitions, a *randomized scheduler* generates $p_i \in \mathbb{R}_{[0,1]}$ for $i = 1, \ldots, n$ such that $\sum_{i=1}^{n} p_i \leq 1$ and then selects transition i with probability p_i or stops with probability $1 - \sum_{i=1}^{n} p_i$. A deterministic scheduler is a special case in which $p_i = 1$ for some i or $p_i = 0$ for each i.

Randomized schedulers, proposed in [22] and applied to the definition of probabilistic trace [23] and testing [24] semantics, probabilistically combine transitions of the original model. Therefore, the resulting resolutions are not necessarily isomorphic to submodels of the original model (or of its unfolding) because a *modification* of the structure of the original model may have taken place. An example of this phenomenon is shown in Fig. 2, where the NPLTS in the leftmost part admits under randomized schedulers the three maximal resolutions depicted next to it in the figure. The resolution starting with z_3 is obtained by combining the two a-transitions departing from s with probabilities p and $1 - p$.

The formalization via a correspondence function of a resolution stemming from a randomized scheduler is not an easy task. The reason is that, according to [22], a combined transition may derive from several *differently labeled* transitions, as shown in the central part of the forthcoming Fig. 3. In other words, a resolution of a simple probabilistic automaton [22], in which every transition has a single label, may have a transition with *several* labels, thereby deviating from a simple probabilistic automaton and hence from an NPLTS.

Similar to [3], below we formalize a resolution induced by a variant of randomized scheduler consistent with the definition of probabilistic bisimilarity given in [25] for simple probabilistic automata. At the current state, the scheduler decides to stop or to perform a certain action among the available ones; in the latter case, it takes a convex combination (i.e., the sum of the values p_i is 1) of the outgoing transitions *identically labeled* with that action. To compensate for the impossibility of combining differently labeled transitions, we admit self-combinations; e.g., in Fig. 3 a combination of the a-transition departing from s with itself n times is able to reproduce the situation in the rightmost part of the same figure, which is equivalent to the one in the central part.

Definition 4. *Let $\mathcal{L} = (S, A, \longrightarrow)$ be an NPLTS and $s \in S$. An acyclic NPLTS $\mathcal{Z} = (Z, A, \longrightarrow_{\mathcal{Z}})$ is a* structure-modifying resolution via randomization *of s, written $\mathcal{Z} \in Res_{\mathrm{sm,r}}(s)$, iff there exists a correspondence function $corr_{\mathcal{Z}} : Z \to S$ such that $s = corr_{\mathcal{Z}}(z_s)$, for some $z_s \in Z$, and for all $z \in Z$ it holds that:*

- *If $z \stackrel{a}{\longrightarrow}_{\mathcal{Z}} \Delta$ then there exist $n \in \mathbb{N}_{\geq 1}$, $p_i \in \mathbb{R}_{]0,1]}$ for $1 \leq i \leq n$ summing up to 1, and $corr_{\mathcal{Z}}(z) \stackrel{a}{\longrightarrow} \Gamma_i$ for $1 \leq i \leq n$, with $corr_{\mathcal{Z}}$ being injective when considered from $supp(\Delta)$ to the disjoint union of the sets $supp(\Gamma_i)$ and satisfying $\Delta(z') = \sum_{i=1}^{n} p_i \cdot \Gamma_i(corr_{\mathcal{Z}}(z'))$ for all $z' \in supp(\Delta)$.*
- *At most one transition departs from z.* ■

Injectivity cannot be directly imposed as in Definition 3, otherwise in Fig. 2 the NPLTS model starting with z_3 would not be a legal resolution induced by the self-combination of the a-transition departing from s' in the rightmost part, and hence s' would not be considered trace equivalent to s in the leftmost part.

3.3   Structure-Modifying Resolutions via Interpolation

For every state in the support of the target distribution of the current transition, an *interpolating scheduler* splits it into $n \in \mathbb{N}_{\geq 1}$ copies, each having a single outgoing transition or no transitions at all, to which probabilities are assigned whose sum is the overall probability of the original state, and then selects one of the copies based on its probability. A deterministic scheduler is a special case in which $n = 1$.

Interpolating schedulers, proposed in [10], probabilistically split states of the original model thereby inducing resolutions possibly modifying the structure of the original model. As mentioned in [10], for each resolution obtained from an

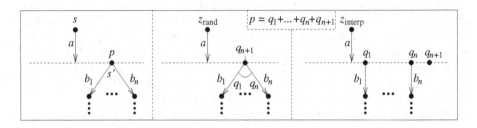

Fig. 3. Equivalent resolutions induced by randomized and interpolating schedulers

interpolating (resp. randomized) scheduler, there exists a resolution obtained from a randomized (resp. interpolating) scheduler with the same trace distribution. This can be seen in Fig. 3, where in the leftmost part we have a state s' reached with probability p in the target distribution of an a-transition. The resolution in the central part, induced by a randomized scheduler that combines the transitions departing from s', is equivalent to the resolution in the rightmost part, induced by an interpolating scheduler that splits state s', where $\sum_{i=1}^{n+1} q_i = p$.

Resolutions arising from interpolating schedulers were natively defined in [10] through a correspondence function that maps all split states to the original state from which they derive. Unlike Definitions 3 and 4, the constraint on transitions is formulated with respect to the states in the support of the corresponding transition of the *original model* – rather than the states in the support of the transition of the resolution – and the preservation of the overall probability associated with each such state makes injectivity requirements unnecessary.

Definition 5. *Let* $\mathcal{L} = (S, A, \longrightarrow)$ *be an NPLTS and* $s \in S$. *An acyclic NPLTS* $\mathcal{Z} = (Z, A, \longrightarrow_\mathcal{Z})$ *is a* structure-modifying resolution via interpolation *of* s, *written* $\mathcal{Z} \in Res_{sm,i}(s)$, *iff there exists a correspondence function* $corr_\mathcal{Z} : Z \to S$ *such that* $s = corr_\mathcal{Z}(z_s)$, *for some* $z_s \in Z$, *and for all* $z \in Z$ *it holds that:*

- *If* $z \xrightarrow{a}_\mathcal{Z} \Delta$ *then* $corr_\mathcal{Z}(z) \xrightarrow{a} \Gamma$, *with* $corr_\mathcal{Z}$ *satisfying for all* $s \in supp(\Gamma)$
  $\Gamma(s) = \sum_{z' \in supp(\Delta)}^{corr_\mathcal{Z}(z')=s} \Delta(z')$.
- *At most one transition departs from* z. ∎

A variant of the structure-modifying resolution above has been proposed in [6], which combines the effect of interpolating and randomized schedulers.

4 Consequences of the Excessive Power of Schedulers

Although deterministic schedulers are very intuitive, the rigid preservation they ensure about the structure of the original model, together with their freedom of performing choices inconsistent with each other in states with equivalent continuations, causes the resulting probabilistic trace equivalence to be overdiscriminating, thereby violating certain desirable properties. This also happens, to a much lesser extent, with randomized and interpolating schedulers. In the following, after presenting in Sect. 4.1 the definition of some probabilistic behavioral equivalences, we illustrate in Sect. 4.2 a number of anomalies.

4.1 Equivalences for Nondeterministic and Probabilistic Processes

The spectrum of behavioral equivalences for nondeterministic and probabilistic processes was studied in [4]. Here we focus on the two endpoints of the spectrum by recalling the definitions of bisimulation and trace semantics.

Probabilistic bisimilarity requires that two NPLTS models are able to mimic each other behavior stepwise, in terms of the probability of reaching the same class of equivalent states when executing the same action [20,25]. Its definition does not need to explicitly resort to resolutions, as these are implicitly built while selecting a single transition from each pair of states.

Definition 6. *Let $(S, A, \longrightarrow)$ be an NPLTS and $s_1, s_2 \in S$. We write $s_1 \sim_{\mathrm{PB}} s_2$ iff there exists a probabilistic bisimulation $\mathcal{B}$ over S such that $(s_1, s_2) \in \mathcal{B}$. An equivalence relation $\mathcal{B}$ over S is a probabilistic bisimulation iff, whenever $(s_1, s_2) \in \mathcal{B}$, then for all $a \in A$ it holds that for each $s_1 \overset{a}{\longrightarrow} \Delta_1$ there exists $s_2 \overset{a}{\longrightarrow} \Delta_2$ such that for all equivalence classes $C \in S/\mathcal{B}$:*

$$\Delta_1(C) = \Delta_2(C) \qquad \blacksquare$$

In contrast, trace equivalence requires that two NPLTS models possess the same trace distributions, i.e., the same family of sets of action sequences weighted with their execution probabilities, where each set is related to a specific resolution of nondeterminism [23]. Its definition, which abstracts from branching points of process behavior, explicitly relies on $Res(\_)$, with which we denote any of the sets of resolutions introduced in Definitions 3 to 5.

Definition 7. *Let $(S, A, \longrightarrow)$ be an NPLTS and $s_1, s_2 \in S$. We write $s_1 \sim_{\mathrm{PTr}} s_2$ iff for each $\mathcal{Z}_1 \in Res(s_1)$ there exists $\mathcal{Z}_2 \in Res(s_2)$ such that for all traces $\alpha \in A^*$:*

$$prob(\mathcal{CC}(z_{s_1}, \alpha)) = prob(\mathcal{CC}(z_{s_2}, \alpha))$$

and also the condition obtained by exchanging $\mathcal{Z}_1$ with $\mathcal{Z}_2$ is satisfied. $\qquad \blacksquare$

4.2   Anomalies and Counterexamples

We now present a number of counterexamples showing that:

- $\sim_{\mathrm{PTr}}$ is not coarser than $\sim_{\mathrm{PB}}$ under deterministic schedulers.
- $\sim_{\mathrm{PTr}}$ is not a congruence w.r.t. action prefix under deterministic schedulers.
- $\sim_{\mathrm{PTr}}$ is not backward compatible with its version for fully prob. processes.

Consider the two NPLTS models in the leftmost part of Fig. 4. It holds that $s_1 \sim_{\mathrm{PB}} s_2$, but $s_1 \not\sim_{\mathrm{PTr}} s_2$ because of the resolution in the central part of Fig. 4, where trace $a\,b$ is executable with probability p instead of 1. This resolution belongs to $Res_{\mathrm{sp}}(s_2) \setminus Res_{\mathrm{sp}}(s_1)$ as it does not preserve the structure of the NPLTS whose initial state is s_1. Notice that the same resolution belongs to $Res_{\mathrm{sm,r}}(s_1)$, if the a-transition of s_1 is combined with itself, and to $Res_{\mathrm{sm,i}}(s_1)$, if z_2' and z_2'' are both mapped to s_1'.

One may be tempted to admit only *maximal* resolutions in the definition of probabilistic trace equivalences, but the problem would still be there if a c-transition departed from z_2''. Moreover, by so doing, probabilistic trace equivalences would no longer be compatible with trace equivalence. For instance, the former would not identify the two fully nondeterministic, trace equivalent NPLTS models in Fig. 4 whose initial states are s_1 and s, because the maximal resolution of s with an a-transition only – featuring traces ε and a – is not matched by the two maximal resolutions of s_1 – resp. featuring also $a\,b$ and $a\,c$.

Let us move to examine the two NPLTS models in the leftmost part of Fig. 5. After the two a-transitions, two distributions are reached that are probabilistic trace equivalent, in the sense that for each class of equivalent states they both assign the same probability to that class. However, it holds that $s_3 \not\sim_{\mathrm{PTr}} s_4$ due to the resolution in the rightmost part of Fig. 5, where trace $a\,a'\,b$ is executable with probability p instead of 1. This resolution belongs to $Res_{\mathrm{sp}}(s_3) \setminus Res_{\mathrm{sp}}(s_4)$ as it does not preserve the structure of the NPLTS whose initial state is s_4. The same resolution belongs to $Res_{\mathrm{sm,r}}(s_4)$, if the a-transition of s_4 is combined with itself, and to $Res_{\mathrm{sm,i}}(s_4)$, if z_3' and z_3'' are both mapped to s_4'.

This example reveals that, under deterministic schedulers, probabilistic trace equivalence is not a congruence with respect to the action prefix operator, which concatenates the execution of an action with a process. The difference with trace equivalence for fully nondeterministic processes is that in our setting the continuation after an action is *not a single process*, but a probability distribution over processes. The problem arises when several equivalent states are in the support of the same distribution, as in the target distribution of the a-transition of s_3, thereby allowing schedulers to act inconsistently.

We finally study the two NPLTS models in the leftmost part of Fig. 6. They are identified by the trace equivalence for fully probabilistic processes of [17], which does not use schedulers as in those processes there are no nondeterministic choices to be solved. However, it turns out that $s_5 \not\sim_{\mathrm{PTr}} s_6$ because $\sim_{\mathrm{PTr}}$ does make use of schedulers, in particular their capability of *stopping the execution*. This is witnessed by the resolution in the rightmost part of Fig. 6, where not only trace $a\,b\,c_1$ but also trace $a\,b$ is executable with probability p. This resolution belongs only to $Res_{\mathrm{sp}}(s_6)$ as it does not preserve the structure of the NPLTS whose initial state is s_5. It does not even belong to $Res_{\mathrm{sm,r}}(s_5) \cup Res_{\mathrm{sm,i}}(s_5)$ because in the NPLTS starting with s_5, after performing the a-transition and the b-transition, the c_1-transition can be executed with probability p, while the c_1-transition in the resolution can be executed with probability 1 and hence its source state cannot be mapped to the source state of the former c_1-transition.

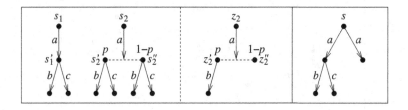

Fig. 4. Violation of $s_1 \sim_{\mathrm{PB}} s_2 \implies s_1 \sim_{\mathrm{PTr}} s_2$ (maximality does not help)

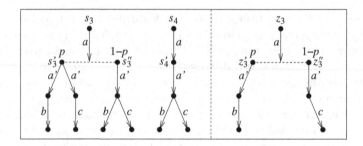

Fig. 5. Violation of congruence with respect to action prefix: $s_3 \not\sim_{\mathrm{PTr}} s_4$

This further example highlights that schedulers inducing structure-modifying resolutions are not exempt from shortcomings despite their greater flexibility. The considered resolution would be ruled out by imposing maximality but, as we have seen at the beginning of this section, that may generate other anomalies.

5 Anomaly Avoidance via Coherent Resolutions

The anomalies shown in Figs. 4, 5 and 6 are due to the freedom of schedulers of making different decisions in equivalent states and cause probabilistic trace equivalence to be overdiscriminating. We thus propose to limit the excessive power of schedulers by restricting them to yield *coherent resolutions*. This means that, if several states in the support of the target distribution of a transition are equivalent, then the decisions made by the scheduler in those states have to be coherent with each other, so that the states to which they correspond in any resolution are equivalent as well. The *coherency constraint* implementing this idea will be expressed by reasoning on *coherent trace distributions*, i.e., families of sets of traces weighted with their execution probabilities in a given resolution, built through the following operations.

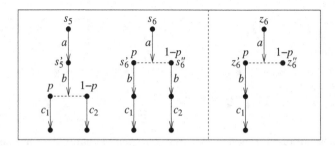

Fig. 6. Incompatibility w.r.t. fully prob. processes: $s_5 \not\sim_{\mathrm{PTr}} s_6$ (levelwise coherency)

Definition 8. *Let $A \neq \emptyset$ be a countable set. For $a \in A$, $p \in \mathbb{R}$, $TD \subseteq 2^{A^* \times \mathbb{R}}$, and $T \subseteq A^* \times \mathbb{R}$ we define:*

$$a \cdot TD = \{a \cdot T \mid T \in TD\} \qquad a \cdot T = \{(a\,\alpha, p') \mid (\alpha, p') \in T\}$$
$$p \cdot TD = \{p \cdot T \mid T \in TD\} \qquad p \cdot T = \{(\alpha, p \cdot p') \mid (\alpha, p') \in T\}$$
$$tr(TD) = \{tr(T) \mid T \in TD\} \quad tr(T) = \{\alpha \in A^* \mid (\alpha, p') \in T \text{ for some } p' \in \mathbb{R}\}$$

while for $TD_1, TD_2 \subseteq 2^{A^* \times \mathbb{R}}$ we define:

$$TD_1 + TD_2 = \begin{cases} \{T_1 + T_2 \mid T_1 \in TD_1 \wedge T_2 \in TD_2 \wedge tr(T_1) = tr(T_2)\} \\ \qquad\qquad\qquad\qquad\qquad \text{if } tr(TD_1) = tr(TD_2) \\ \{T_1 + T_2 \mid T_1 \in TD_1 \wedge T_2 \in TD_2\} \\ \qquad\qquad\qquad\qquad\qquad \text{otherwise} \end{cases}$$

where for $T_1, T_2 \subseteq A^* \times \mathbb{R}$ we define:

$$T_1 + T_2 = \{(\alpha, p_1 + p_2) \mid (\alpha, p_1) \in T_1 \wedge (\alpha, p_2) \in T_2\} \cup$$
$$\{(\alpha, p) \in T_1 \cup T_2 \mid \alpha \notin tr(T_1) \cap tr(T_2)\} \qquad \blacksquare$$

Weighted trace set addition is commutative and associative. In the definition of $T_1 + T_2$, which is inspired by [3], probabilities of identical traces in the two summands are *always* added up for coherency purposes. Before Definition 3.5 of [3], the definition of $X + Y$, i.e., $T_1 + T_2$, should have included $(\alpha, q) \in X \cup Y$ in the sum anyhow, otherwise the right-to-left implication in Lemma 3.7 of [3] cannot hold as can be seen from trace $a\,b$ of the (incoherent) resolution in the central part of Fig. 4 of this paper; that definition of $X + Y$ works here instead, because of the focus on coherency.

Trace distribution addition is only commutative. Intuitively, the two summands in $TD_1 + TD_2$ represent two families of sets of weighted traces executable in the resolutions of two states in the support of a target distribution. Every weighted trace set $T_1 \in TD_1$ is summed with every weighted trace set $T_2 \in TD_2$ – so to characterize an overall resolution – unless TD_1 and TD_2 have the same family of trace sets, in which case summation is restricted to weighted trace sets featuring the same traces for the sake of coherency. In the definition below, the double summation ensures that trace distributions $\Delta(s') \cdot TD^c_{n-1}(s')$ exhibiting the same family Θ of trace sets will be summed up first.

Definition 9. *Let $\mathcal{L} = (S, A, \longrightarrow)$ be an NPLTS and $s \in S$. The coherent trace distribution of s is the subset of $2^{A^* \times \mathbb{R}_{]0,1]}}$ defined as follows:*

$$TD^c(s) = \bigcup_{n \in \mathbb{N}} TD^c_n(s)$$

where the coherent trace distribution of s whose traces have length at most n is defined as:

$$TD_n^c(s) = \begin{cases} (\varepsilon,1) \dagger \bigcup_{s \xrightarrow{a} \Delta} a \cdot \left(\sum_{\Theta \in tr(\Delta,n-1)} \overset{tr(TD_{n-1}^c(s'))=\Theta}{\sum_{s' \in supp(\Delta)}} \Delta(s') \cdot TD_{n-1}^c(s') \right) \\ \qquad\qquad\qquad\qquad\qquad\qquad \textit{if } n > 0 \textit{ and } s \textit{ has outgoing transitions} \\ \{\{(\varepsilon,1)\}\} \\ \qquad\qquad\qquad\qquad\qquad\qquad \textit{otherwise} \end{cases}$$

for $tr(\Delta, n-1) = \{tr(TD_{n-1}^c(s')) \mid s' \in supp(\Delta)\}$ and $(\varepsilon,1) \dagger TD = \{\{(\varepsilon,1)\} \cup T \mid T \in TD\}$. ∎

Let us reconsider the three counterexamples of Sect. 4 plus two more:

- In Fig. 4 we have $TD^c(s_2') = \{\{(\varepsilon,1)\}, \{(\varepsilon,1), (b,1)\}, \{(\varepsilon,1), (c,1)\}\} = TD^c(s_2'')$ – from which $TD^c(s_2) = \{\{(\varepsilon,1)\}, \{(\varepsilon,1), (a,1)\}, \{(\varepsilon,1), (a,1), (a\,b,1)\}, \{(\varepsilon,1), (a,1), (a\,c,1)\}\} = TD^c(s_1)$ follows – but in the resolution $TD^c(z_2') = \{\{(\varepsilon,1)\}, \{(\varepsilon,1), (b,1)\}\} \neq \{\{(\varepsilon,1)\}\} = TD^c(z_2'')$.
- In Fig. 5 we have $TD^c(s_3') = \{\{(\varepsilon,1)\}, \{(\varepsilon,1), (a',1)\}, \{(\varepsilon,1), (a',1), (a'\,b,1)\}, \{(\varepsilon,1), (a',1), (a'\,c,1)\}\} = TD^c(s_3'')$ whereas in the resolution $TD^c(z_3') = \{\{(\varepsilon,1)\}, \{(\varepsilon,1), (a',1)\}, \{(\varepsilon,1), (a',1), (a'\,b,1)\}\} \neq \{\{(\varepsilon,1)\}, \{(\varepsilon,1), (a',1)\}, \{(\varepsilon,1), (a',1), (a'\,c,1)\}\} = TD^c(z_3'')$.
- In Fig. 6 we have $TD^c(s_6') = \{\{(\varepsilon,1)\}, \{(\varepsilon,1), (b,1)\}, \{(\varepsilon,1), (b,1), (b\,c_1,1)\}\} \neq \{\{(\varepsilon,1)\}, \{(\varepsilon,1), (b,1)\}, \{(\varepsilon,1), (b,1), (b\,c_2,1)\}\} = TD^c(s_6'')$. However, $TD_1^c(s_6') = \{\{(\varepsilon,1), (b,1)\}\} = TD_1^c(s_6'')$ while in the resolution $TD_1^c(z_6') = \{\{(\varepsilon,1), (b,1)\}\} \neq \{\{(\varepsilon,1)\}\} = TD_1^c(z_6'')$. This shows that we should set up separate coherency constraints relying on TD_n^c sets for every $n \in \mathbb{N}$.
- Consider the two fully probabilistic NPLTS models in the leftmost part of Fig. 7. They are identified by the trace equivalence of [17], but $s_7 \not\sim_{\text{PTr}} s_8$ due to the resolution in the rightmost part of the same figure. It holds that $TD_2^c(s_7') = \{\{(\varepsilon,1), (b,1), (b\,c,0.3)\}\} \neq \{\{(\varepsilon,1), (b,1), (b\,c,0.2)\}\} = TD_2^c(s_7'')$, with $TD_3^c(s_7) = \{\{(\varepsilon,1), (a,1), (a\,b,1), (a\,b\,c,0.25)\}\} = TD_3^c(s_8)$. However, we observe that $tr(TD_2^c(s_7')) = \{\{\varepsilon, b, b\,c\}\} = tr(TD_2^c(s_7''))$ whereas $tr(TD_2^c(z_7')) = \{\{\varepsilon, b, b\,c\}\} \neq \{\{\varepsilon, b\}\} = tr(TD_2^c(z_7''))$. This indicates that the coherency constraints should rely on TD_n^c sets up to the probabilities they contain, i.e., the coherency constraints should rely on $tr(TD_n^c)$ sets.
- The violations in Figs. 6 and 7 of backward compatibility with the trace equivalence of [17] have a twofold interpretation. The former is that incoherent selections are made by the scheduler in states having the same traces of a certain length. The latter ascribes the lack of coherency to the fact that, in both resolutions depicted in those figures, the scheduler proceeds by selecting a transition along one direction while it stops the execution along the other direction. This is even more evident with the two fully probabilistic NPLTS models in the leftmost part of Fig. 8, which are identified by [17] but told apart by the resolution on the right, where $a\,b\,c_1$ is executable with probability 0.25, as $tr(TD_2^c(s_{10}'))$, $tr(TD_2^c(s_{10}''))$, and $tr(TD_2^c(s_{10}'''))$ are pairwise

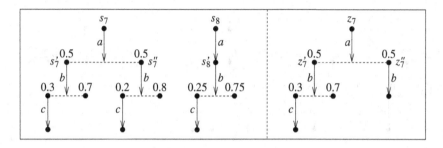

Fig. 7. Incompatibility w.r.t. fully prob. processes: $s_7 \nsim_{\mathrm{PTr}} s_8$ (probability abstraction)

different. In every coherent resolution of s_9, trace $a\,b\,c_1$ can be executed only with probability 0.5. This calls for a complete presence of computations of the same length in each resolution – including shorter maximal computations if any – which is different from requiring resolution maximality.

Definition 10. *Let* $\mathcal{L} = (S, A, \longrightarrow)$ *be an NPLTS,* $s \in S$, *and* $\mathcal{Z} = (Z, A, \longrightarrow_{\mathcal{Z}}) \in Res(s)$ *with correspondence function* $corr_{\mathcal{Z}} : Z \to S$. *We say that* $\mathcal{Z}$ *is a* coherent resolution *of* s, *written* $\mathcal{Z} \in Res^{c}(s)$, *iff for all* $z \in Z$, *whenever* $z \overset{a}{\longrightarrow}_{\mathcal{Z}} \Delta$, *then for all* $n \in \mathbb{N}$:

1. $tr(TD^{c}_{n}(corr_{\mathcal{Z}}(z'))) = tr(TD^{c}_{n}(corr_{\mathcal{Z}}(z''))) \implies tr(TD^{c}_{n}(z')) = tr(TD^{c}_{n}(z''))$
   for all $z', z'' \in supp(\Delta)$.
2. *If there exists* $z' \in supp(\Delta)$ *such that* $tr(TD^{c}_{n}(z'))$ *contains traces of length* n, *then for all* $z'' \in supp(\Delta)$ *either* $tr(TD^{c}_{n}(z''))$ *contains traces of length* n *too, or any* $\alpha \in A^{*}$ *occurring in* $tr(TD^{c}_{n}(z''))$ *has length less than* n *but there exists a maximal trace in* $tr(TD^{c}_{n}(corr_{\mathcal{Z}}(z'')))$ *corresponding to* α. ∎

In the definition above, $Res(\_)$ denotes any of the sets of resolutions introduced in Definitions 3 to 5. From now on, we focus on $Res^{c}_{sp}(\_)$. Notice that the resolutions in Figs. 4 to 8 do *not* respectively belong to $Res^{c}_{sp}(s_2)$, $Res^{c}_{sp}(s_3)$, $Res^{c}_{sp}(s_6)$, $Res^{c}_{sp}(s_7)$, and $Res^{c}_{sp}(s_{10})$.

We conclude by proving that probabilistic trace equivalence no longer suffers from the anomalies illustrated in Sect. 4 *when using coherent resolutions* induced by deterministic schedulers. In the following, we lift a probabilistic behavioral equivalence $\sim$ from states to distributions over states by letting $\Delta_1 \sim \Delta_2$ iff $\Delta_1(C) = \Delta_2(C)$ for all equivalence classes C of $\sim$. Moreover, the action prefix construction $a\,.\,\Delta$ stands for an a-transition whose target distribution is Δ, whereas $\sim^{\mathrm{fp}}_{\mathrm{PTr}}$ denotes the probabilistic trace equivalence for fully probabilistic processes defined in [17] by letting $s_1 \sim^{\mathrm{fp}}_{\mathrm{PTr}} s_2$ iff $prob(\mathcal{CC}(s_1, \alpha)) = prob(\mathcal{CC}(s_2, \alpha))$ for all $\alpha \in A^{*}$.

We point out that coherency was unfortunately neglected in [3,4]. In particular, property 1 below is the rectified version of a chain of results in [4] consisting of Thms. 6.5(2), 5.9(3), 4.5(2) and property 3 below is the rectified version

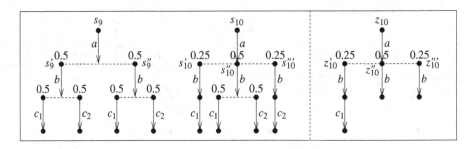

Fig. 8. Incompat. w.r.t. fully prob. processes: $s_9 \nsim_{\mathrm{PTr}} s_{10}$ (levelwise completeness)

of Thm. 3.4(2) of [3,4]; deterministic schedulers were considered in all those theorems. Property 3 now holds also in the case of randomized/interpolating schedulers by just imposing condition 2 of Definition 10.

Theorem 1. *Let $\mathcal{L} = (S, A, \longrightarrow)$ be an NPLTS, $s_1, s_2 \in S$, $\Delta_1, \Delta_2 \in Distr(S)$. Under coherent resolutions induced by deterministic schedulers it holds that:*

1. $s_1 \sim_{\mathrm{PB}} s_2 \implies s_1 \sim_{\mathrm{PTr}} s_2$.
2. $\Delta_1 \sim_{\mathrm{PTr}} \Delta_2 \implies a \,.\, \Delta_1 \sim_{\mathrm{PTr}} a \,.\, \Delta_2$ *for all $a \in A$.*
3. *If $\mathcal{L}$ is fully probabilistic, then $s_1 \sim_{\mathrm{PTr}} s_2 \iff s_1 \sim_{\mathrm{PTr}}^{\mathrm{fp}} s_2$.* ∎

We finally observe that looser coherency constraints, based on weighted trace sets rather than trace distributions as in Definition 10, would not work. Similar to $TD^c(s)$ in Definition 9, one may define $T^c(s)$ by considering all weighted traces executable from s at once – i.e., without keeping track of the resolutions in which they are feasible – and use it for coherency purposes, but then probabilistic trace equivalent NPLTS models like the ones in Fig. 9 would be told apart. Indeed, we would have $tr(T^c(s_1')) = \{\epsilon, b, b\,c_1, b\,c_2, b\,c\} = tr(T^c(s_2'))$ – whereas $tr(TD^c(s_1')) \neq tr(TD^c(s_2'))$ – hence in any coherent resolution of s' traces $a\,b\,c_1$, $a\,b\,c_2$, $a\,b\,c$ could only be executed with probability 0.5 if present, while s'' admits coherent resolutions in which those traces have execution probability 0.25.

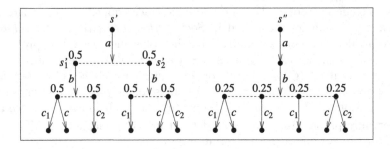

Fig. 9. Using weighted trace sets for coherency breaks probabilistic trace equivalence

6 Conclusions

To guarantee a number of desirable properties for probabilistic trace equivalence over probabilistic automata, we have proposed a set of coherency constraints as a solution to the problem – addressed also in [12] for a different probabilistic model and equivalence – of limiting the excessive power of schedulers.

The highlighted anomalies mostly have to do with structure-preserving resolutions generated by deterministic schedulers, so one may wonder why not to avoid those schedulers altogether. The first reason is that, as shown in [4], the use of a specific family of schedulers has an impact on the discriminating power of behavioral equivalences, so there might be situations in which considering deterministic schedulers is more appropriate. The second reason is that, as witnessed by Fig. 6, some of the examined anomalies affect also equivalences defined on structure-modifying resolutions generated by randomized/interpolating schedulers. The third reason is that in more general frameworks, like the ULTraS metamodel [2] of which probabilistic automata are an instance, the applicability of deterministic schedulers is always possible, while this might not be the case for other families of schedulers.

Acknowledgement. We would like to thank Valeria Vignudelli for pointing out the property violation illustrated in Fig. 4 and Rob van Glabbeek for the valuable discussions on interpolating and randomized schedulers.

References

1. Baier, C., Katoen, J.-P., Hermanns, H., Wolf, V.: Comparative branching-time semantics for Markov chains. Inf. Comput. **200**, 149–214 (2005)
2. Bernardo, M.: Genesis and evolution of ULTraS: metamodel, metaequivalences, metaresults. In: Boreale, M., Corradini, F., Loreti, M., Pugliese, R. (eds.) Models, Languages, and Tools for Concurrent and Distributed Programming. LNCS, vol. 11665, pp. 92–111. Springer, Cham (2019). https://doi.org/10.1007/978-3-030-21485-2_7
3. Bernardo, M., De Nicola, R., Loreti, M.: Revisiting trace and testing equivalences for nondeterministic and probabilistic processes. Logical Methods Comput. Sci. **10**(1:16), 1–42 (2014)
4. Bernardo, M., De Nicola, R., Loreti, M.: Relating strong behavioral equivalences for processes with nondeterminism and probabilities. Theor. Comput. Sci. **546**, 63–92 (2014)
5. Bernardo, M., Sangiorgi, D., Vignudelli, V.: On the discriminating power of testing equivalences for reactive probabilistic systems: results and open problems. In: Norman, G., Sanders, W. (eds.) QEST 2014. LNCS, vol. 8657, pp. 281–296. Springer, Cham (2014). https://doi.org/10.1007/978-3-319-10696-0_23
6. Bonchi, F., Sokolova, A., Vignudelli, V.: The theory of traces for systems with nondeterminism and probability. In: Proceedings of the 34th ACM/IEEE Symposium on Logic in Computer Science (LICS 2019), no. (19:62), pp. 1–14. IEEE-CS Press (2019)
7. Brookes, S.D., Hoare, C.A.R., Roscoe, A.W.: A theory of communicating sequential processes. J. ACM **31**, 560–599 (1984)

8. De Nicola, R.: Extensional equivalences for transition systems. Acta Informatica **24**, 211–237 (1987)

9. De Nicola, R., Hennessy, M.: Testing equivalences for processes. Theor. Comput. Sci. **34**, 83–133 (1984)

10. Deng, Y., van Glabbeek, R., Morgan, C., Zhang, C.: Scalar outcomes suffice for finitary probabilistic testing. In: De Nicola, R. (ed.) ESOP 2007. LNCS, vol. 4421, pp. 363–378. Springer, Heidelberg (2007). https://doi.org/10.1007/978-3-540-71316-6_25

11. Derman, C.: Finite State Markovian Decision Processes. Academic Press, Cambridge (1970)

12. Georgievska, S., Andova, S.: Probabilistic may/must testing: retaining probabilities by restricted schedulers. Formal Aspects Comput. **24**, 727–748 (2012)

13. van Glabbeek, R.J.: The linear time - branching time spectrum I. In: Handbook of Process Algebra, pp. 3–99. Elsevier (2001)

14. Huynh, D.T., Tian, L.: On some equivalence relations for probabilistic processes. Fundamenta Informaticae **17**, 211–234 (1992)

15. Jonsson, B., Ho-Stuart, C., Yi, W.: Testing and refinement for nondeterministic and probabilistic processes. In: Langmaack, H., de Roever, W.-P., Vytopil, J. (eds.) FTRTFT 1994. LNCS, vol. 863, pp. 418–430. Springer, Heidelberg (1994). https://doi.org/10.1007/3-540-58468-4_176

16. Jonsson, B., Yi, W.: Compositional testing preorders for probabilistic processes. In: Proceedings of the 10th IEEE Symposium on Logic in Computer Science (LICS 1995), pp. 431–441. IEEE-CS Press (1995)

17. Jou, C.-C., Smolka, S.A.: Equivalences, congruences, and complete axiomatizations for probabilistic processes. In: Baeten, J.C.M., Klop, J.W. (eds.) CONCUR 1990. LNCS, vol. 458, pp. 367–383. Springer, Heidelberg (1990). https://doi.org/10.1007/BFb0039071

18. Keller, R.M.: Formal verification of parallel programs. Commun. ACM **19**, 371–384 (1976)

19. Kemeny, J.G., Snell, J.L.: Finite Markov Chains. Van Nostrand, New York (1960)

20. Larsen, K.G., Skou, A.: Bisimulation through probabilistic testing. Inf. Comput. **94**, 1–28 (1991)

21. Milner, R.: Communication and Concurrency. Prentice Hall, Upper Saddle River (1989)

22. Segala, R.: Modeling and verification of randomized distributed real-time systems. Ph.D. thesis (1995)

23. Segala, R.: A compositional trace-based semantics for probabilistic automata. In: Lee, I., Smolka, S.A. (eds.) CONCUR 1995. LNCS, vol. 962, pp. 234–248. Springer, Heidelberg (1995). https://doi.org/10.1007/3-540-60218-6_17

24. Segala, R.: Testing probabilistic automata. In: Montanari, U., Sassone, V. (eds.) CONCUR 1996. LNCS, vol. 1119, pp. 299–314. Springer, Heidelberg (1996). https://doi.org/10.1007/3-540-61604-7_62

25. Segala, R., Lynch, N.: Probabilistic simulations for probabilistic processes. In: Jonsson, B., Parrow, J. (eds.) CONCUR 1994. LNCS, vol. 836, pp. 481–496. Springer, Heidelberg (1994). https://doi.org/10.1007/978-3-540-48654-1_35

26. Yi, W., Larsen, K.G.: Testing probabilistic and nondeterministic processes. In: Proceedings of the 12th International Symposium on Protocol Specification, Testing and Verification (PSTV 1992), pp. 47–61. North-Holland (1992)

Emulating Self-adaptive Stochastic Petri Nets

Lorenzo Capra[1] and Matteo Camilli[2]([✉])

[1] Department of Computer Science, Università degli Studi di Milano, Milan, Italy
`lorenzo.capra@unimi.it`
[2] Faculty of Computer Science, Free University of Bozen-Bolzano, Bolzano, Italy
`matteo.camilli@unibz.it`

Abstract. Traditional Petri nets lack specific features to conveniently describe systems with an evolving structure. A model based on the Symmetric Net formalism has been recently introduced. It is composed of an emulator reproducing the behaviour of a Place/Transition net (encoded as a marking) and a basic set of net-transformation primitives to specify evolutionary behaviour. In this paper, we discuss the adoption of the stochastic extension of Symmetric Nets for performance analysis, considering important issues related to time specification and analysis complexity. We put into place theoretical aspects by using a running example consisting in a self-healing manufacturing system.

Keywords: Evolving systems · Stochastic petri nets · Symmetric nets

1 Introduction

Both low- and high-level Petri nets (PNs) lack features to describe in a simple way structural changes that may occur in a wide class of systems, like reconfigurable, self-adaptive, self-healing, and so forth. To bridge this gap, new PN-based formalisms have been proposed in literature in the last decade, often hybrid and/or characterized by complex annotations, in which enhanced modelling capabilities are not supported by analysis techniques and tools.

A formal model for evolving systems based on the Symmetric Nets (SNs) formalism (formerly known as Well-formed Nets) [8] has been recently introduced in [6]. The idea takes inspiration from the "nets within nets" paradigm [12,15] and is based on a meta-level net (a SN) emulating a system-level net (a Place/-Transition -P/T- system with inhibitor arcs, known to be Turing-complete) encoded as a marking of the meta-net. The modeling approach supplies a set of *transformation primitives* (SN subnets), accessible through a simple API, which can be profitably used to specify adaptation procedures acting on the emulated system. The resulting approach is conceptually uniform and simple, differently from other proposals with similar objectives. SNs are a flavour of *Colored* Petri Nets [10] characterized by a structured syntax implicitly capturing system symmetries, which may be exploited to reduce the complexity of analysis techniques.

M. Gribaudo et al. (Eds.): EPEW 2019, LNCS 12039, pp. 33–49, 2020.
https://doi.org/10.1007/978-3-030-44411-2_3

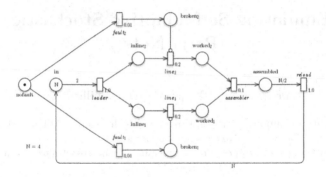

Fig. 1. MS nominal behaviour.

SN were originally proposed in their stochastic version, called SSN. The use of the SN formalism offers the great advantage of exploiting a well engineered, off-the-shelf tool, GREATSPN [1], that natively supports it. In this paper we discuss the usage of the emulation-based model for performance analysis. Relevant issues concerning time specification, time semantics preservation, and analysis complexity, are addressed. A simple manufacturing system (MS) with self-healing capabilities is used as a running example, to illustrate the core concepts and discuss some experimental results. The emulation-based technique is general: any P/T model can be emulated with this approach. The entire emulation process may be easily automated, even if currently only some steps are: the emulator's initial marking is directly derived from a P/T net in PNML format, whereas the ALGEBRA module of GREATSPN is used to link the user-defined adaptation procedures to the emulator. There is ongoing work to set up the model's stochastic parameters in a semi-automated way, as discussed in Sect. 5. Moreover, the design of adaptation sub-nets is completely unaware of emulator inside.

Related Work. The emulator-based approach has been introduced in [5] using pure spec-inscribed PN [10]. Thanks to the abstraction provided by this formalism, the resulting model is much more compact than the one considered here. It has been implemented as an extensible library [4] but, currently, with reduced analysis capabilities (interactive simulation and a LTL model checker are available) and without timing. A survey on approaches combining higher-order tokens and the features of object-orientation can be found in [15]. Reference Nets [2] are the representative of this class of formalisms, and are supported by tool **Renew**. The formalism introduced in [9] extends Algebraic Higher-Order (AHO) nets (i.e., HLPNs annotated with higher-order algebraic language) with the main concepts of graph transformation systems. Common drawbacks of models based on algebraic-functional languages and/or higher-order tokens are the lack of a sound/clear semantics, the use of hybrid formalisms hard to manage by non experts, and, consequently, a limited support in terms of tools/analysis techniques. Considering time extensions, there are a few available models that refer to PNs oriented to real-time systems [3]. The formalism introduced in [14] integrates GSPNs with graph-rewriting rules and supports simple evolution patterns. For a general survey on available formalisms for self-adaptive systems we let the reader refer to [16].

2   Background

Assuming that the reader is familiar with low-level PN, let us give a brief overview of P/T nets [13], exemplified by Fig. 1. In these nets there a 3 kinds of arcs, input/output/inhibitor (the latter drawn with a small ending circle), described by multisets on $P \times T$, i.e., maps $P \times T \to \mathbb{N}$, where P and T are (disjoint, non-empty, and finite) sets holding the net's places and transitions. Null-weight arcs are not drawn. A marking m (i.e., a system state) is a multiset on P. A transition $t \in T$ is enabled in m iff $\forall p, I(p, t) \leq m(p) \wedge (H(p, t) = 0 \vee H(p, t) > m(p))$. If enabled, t may fire leading to m', where $m'(p) = m(p) + O(p, t) - I(p, t)$. This is denoted $m[t > m'$. A P/T system is a P/T net with an initial marking m_0. Its *reachability graph* is a multi-graph whose nodes are the markings reachable from m_0, and such that there is an edge $m \xrightarrow{t} m'$ iff $m[t > m'$.

In SN [8] (see Fig. 3), like in any high-level PN formalism, nodes are associated with domains, expressed as Cartesian products of finite *color classes*. A color class, denoted hereinafter by a capital letter, e.g., C, may be partitioned into *static subclasses* $\{C_j\}$. Colours in a class represent entities of the same nature, but only colors within the same static subclass are guaranteed to *behave similarly*. A color class may also be *circularly ordered*. The SN in Fig. 3 has two basic color classes, $P = \{pl_i\}$ and $T = \{tr_j\}$, encoding the nodes of a P/T net. They may have to be partitioned and/or ordered for the sake of modelling.

A place's color domain, $cd(p)$, defines the type of tokens the place may hold. Places IN, OUT, H, with domain $P \times T$, encode the structure of a P/T net. The domain of place MARK is P, in fact, this place encodes a P/T net marking. Each place contains a multiset defined on its domain, the place's *marking*. For example, the multiplicity of the token $\langle pr_i, tr_k \rangle$ in place IN encodes the weight of a P/T arc from pl_i to tr_k. A SN marking is denoted M, with $M(p) \in Bag[cd(p)]$, where $Bag[D]$ is the set of multisets defined on domain D.

SN transitions represent parametrized events. The *instances* of t are elements of its domain, $cd(t)$, which is implicitly defined by the typed variables annotating the arcs which surround t (assuming an implicit order among them). A variable is denoted by a small letter, which refers to the variable's color class, with a subscript, possibly omitted when there is one variable of a given type. A transition instance, also denoted (t, b), is a binding of t's variables with colors of proper type. Transition nextT in Fig. 3 has got one variable, t, therefore its domain is T. Transition sc_H has got variables p, t_1, t_2, so its domain is $P \times T \times T$.

A *guard* can be used to restrict the domain of t: it is a logical expression defined on $cd(t)$, and its terms, called *basic predicates*, allow one to (1) compare colors assigned to variables of the same type ($c_1 = c_2, c_1 \neq c_2$); (2) test whether a color belongs to a given static subclass ($c_1 \in C_i$); (3) compare the static subclasses of the colors assigned to two variables ($d(c_1) = d(c_2), d(c_1) \neq d(c_2)$).

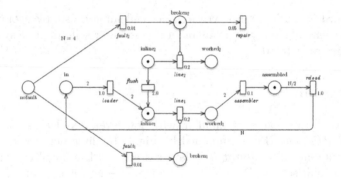

Fig. 2. MS self-adaptation upon a fault on line 2.

Again, there are input/output/inhibitor arcs linking places to transitions, annotated by *functions*, denoted $W^-[p,t]$, $W^+[p,t]$ and $W^h[p,t]$, respectively. An arc function is a map $cd(t) \rightarrow Bag[cd(p)]$, formally expressed as a linear combination of *tuples* $\langle f_1, \ldots, f_k \rangle$ of *class functions*. A class-C function f_i is a map $cd(t) \rightarrow Bag[C]$, expressed in turn as a linear combination of elementary functions chosen among $c_j, c_j{+}{+}, C_q, All$: c_j (variable, or *projection*) maps a tuple in $cd(t)$ to the jth occurrence of color C in it; $++$ gets the successor $mod_{|C|}$, if the class is ordered; C_q and *All* are *constants* mapping to $\sum_{x \in C_q} x$ and $\sum_{x \in C} x$, respectively. The evaluation of $\langle f_1, \ldots, f_k \rangle$ on $b \in cd(t)$ results in $f_1(b) \times \ldots f_k(b)$, where $\times$ is the multiset Cartesian product.

For example, the firing of nextT with binding $t = \text{tr}_1$ replaces color tr_1 with its successor, tr_2, in place toTest, and withdraws tr_2 from AllT. The function $\langle All \rangle$ on the inhibitor arc linking place checkList to transition endTestEnab checks for the absence of tokens in that place. The same function on the output arc from endTestEnab to AllT makes this place be filled with all the colors in T. The tuple $\langle p_1, t_1 + t_2 \rangle$ evaluated on the binding $p_1 = \text{pl}_3, t_1 = \text{tr}_1, t_2 = \text{tr}_4...$, results in the multiset $\langle \text{pl}_3, \text{tr}_1 \rangle + \langle \text{pl}_3, \text{tr}_4 \rangle$. And so forth.

The SN formalism admits two different kinds of transitions: transitions drawn as rectangles represent *observable* (or time-consuming) events, whereas those drawn as tiny black bars represent *invisible* (or logical) activities. The latter take priority over the former, assumed to have priority 0. "Black" transitions may have different priorities, specified by a map $\pi : T_{\text{black}} \rightarrow \mathbb{N}^+$.

(t, b) *has concession* in marking M if $\forall p \quad W^-[p,t](b) \leq M(p) \wedge \forall c \in cd(p)$ $W^h[p,t](b)(c) = 0 \vee W^h[p,t](b)(c) > M(p)(c)$. An instance (t, b) having concession in M is *enabled* if no higher priority transition instance has. In this case it may fire, leading to M', where $\forall p \ M'(p) = M(p) - W^-[p,t](b) + W^+[p,t](b)$. This is denoted $M[(t, b) > M'$. If σ is a sequence of transition instances, $M[\sigma > M'$ means that M' is reachable from M through σ.

A SN marking is called *vanishing* if some black transition is enabled, *tangible* otherwise. Assuming that the initial marking M_0 is tangible, and no cyclic paths of black transitions do exist, we may build the tangible reachability graph (TRG) of a SN, whose nodes are tangible markings such that $M_i \xrightarrow{\beta} M_j$ if and only if $M_i[\beta\sigma > M_j$, where β is an observable transition instance whereas σ is a (possibly null) sequence of black transition instances.

3   A Self-adaptive Manufacturing System Example

As a running example, we use a simple manufacturing system (MS) equipped with self-healing capabilities. The MS is composed of two symmetric production lines working a number of raw pieces which are loaded (two at a time) into the system and evenly distributed to the lines. Pairs of worked pieces are then assembled into the final artifact. Either line is periodically subject to failures (the possibility of simultaneous faults on both lines is considered null). The system's nominal behaviour, and the fault occurrence, are shown in Fig. 1: whenever either place $broken_1$ or $broken_2$ is marked the corresponding line is blocked and the MS, without any further action, would eventually get stuck.

A first adaptation scenario is represented by the MS reconfiguration upon a failure. At the end of adaptation the MS layout looks like Fig. 2. During adaptation the MS is not shut down, but continues working using the available resources: the faulty line is detached, and the behaviour of both the loader and the assembler are changed accordingly. The presence of pending pieces on the faulty line is a critical issue. Once adapted, the loader puts two row pieces at a time on the available line, whereas the assembler takes pairs of worked pieces from that line. The second scenario brings the MS back to its nominal configuration as soon as the faulty component has been repaired. Once again, without stopping the system. Despite its simplicity, trying to model the MS and its adaptation with PN (even if high-level) is very hard and costly. Our approach follows a clear separation of concerns and consists of representing the adaptation *procedures* as apart components (SN sub-nets) running on a distributed infrastructure, which concurrently monitors the current state/topology of a base-level (P/T) system and possibly rearrange it. The base-level system's dynamics is emulated by a meta-level SN model encoding the system as a marking. Adaptation is implemented by read/write primitives (SN sub-nets) which safely operate on the system's encoding, through a simple API. We can thus easily represent disconnection/reconnection of a faulty/repaired production line, migration of raw pieces from one line to the other, failure repair (through a newly created transition), and so forth.

4   The SN-Based Emulating Framework

The SN model in Fig. 3 (called *emulator*) is the building-block of a modelling approach to highly dynamic discrete-event systems based on SN. It reproduces the *interleaving* semantics of a P/T system encoded as a marking. The emulator (and the whole framework based on it) can be interactively simulated and analysed by means of GREATSPN[1]. Despite its complexity, caused by the lack

---

[1] All the SN sources (GREATSPN .PNPRO files) are publicly available at https://github.com/SELab-unimi/sn-based-emulator.

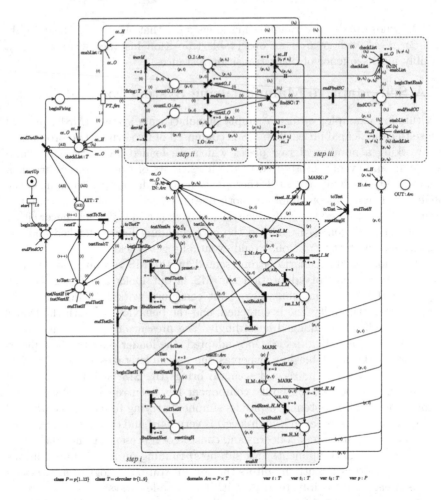

Fig. 3. Symmetric net emulating model.

of abstraction in SN color structure, it allows for significant achievements in dynamic system modelling. The depicted version is one of the three available ones, pretty similar but different in analysis capability (Sect. 7).

4.1   The Emulator Model

The emulator's color annotations build on basic classes P and T, whose elements represent the nodes of a P/T net. For simplicity, we identify colors and corresponding nodes. Let $\mathcal{N} = (P, T, I, O, H, m_0)$ be a P/T system: classes P and T contain the nodes of $\mathcal{N}$(P $\supseteq$ P, T $\supseteq$ T), and should be large enough to cover its possible evolutions. The emulator's places IN, OUT, H (with domain P $\times$ T) and MARK (cd(MARK) = P) encode the (current) structure and marking of

$\mathcal{N}$, respectively. Their initial marking is: $\forall \text{pl}_i \in P, \forall \text{tr}_j \in T$

$$M_0(\text{IN})[\langle \text{pl}_i, \text{tr}_j \rangle)] = I(\text{pl}_i, \text{tr}_j) \qquad M_0(\text{OUT})[\langle \text{pl}_i, \text{tr}_j \rangle)] = O(\text{pl}_i, \text{tr}_j)$$
$$M_0(\text{H})[\langle \text{pl}_i, \text{tr}_j \rangle] = H(\text{pl}_i, \text{tr}_j) \qquad M_0(\text{MARK})[\langle \text{pl}_i \rangle] = m_0(\text{pl}_i).$$

For convenience, the SN places O_I and I_O encode the functions $O-I$ and $I-O$.

The emulator has a cyclic behaviour, which can be summarized as follows (see [6] for further details). Any reachable *tangible* marking corresponds to a reachable marking of the encoded P/T system, any enabled instance of transition PT_fire (the only *observable*, but for start) matches an enabled P/T transition. The firing of (PT_fire, $t = \text{tr}_k$) triggers a sequence σ of black transition instances reproducing the firing of tr_k, according to the *atomic* semantics of PN. The sequence σ is composed of three parts: (*step ii*) the marking of place MARK is updated, according to the P/T firing rule; (*step iii*) two lists holding the transitions that were enabled before the firing of tr_k (place enabList) and those whose enabling must be checked upon it (checkList) are efficiently updated, taking into account the *structural conflict* (SC) and *causal connection* (SCC) relations. Both steps *ii* and *iii* rely on the information held in places O_I and I_O; (*step i*) all and only the transitions marked as "to be checked" are tested for enabling, first considering *input* places then *inhibitor* places. After place checkList has been emptied, and enabList updated, the emulation cycle restarts (transition endTestEnab). The emulator can be initialized by either putting a (neutral) token in place startUp and all transitions of $\mathcal{N}$ in checkList, or a token in beginFiring and the precomputed set of transitions enabled in m_0 in enabList.

For any encoded P/T system $\mathcal{N}$, the following properties holds:

1. $m_i[\text{tr}_k > m_j$ if and only if $M_i[\beta \cdot \sigma > M_j$, where M_i and M_j are emulator's tangible markings such that $M_i(\text{MARK})$ and $M_j(\text{MARK})$ correspond to m_i and m_j, respectively, $\beta = (\text{PT\_fire}, t = \text{tr}_k)$, and σ is a sequence of immediate transition instances
2. if $M_i[\beta \cdot \sigma > M_j$ and $M_i[\beta \cdot \sigma' > M_h$ then $M_j = M_h$.

The *tangible* reachability graph of the emulator encoding $\mathcal{N}$ therefore is *isomorphic* to the reachability graph of $\mathcal{N}$.

4.2   The Evolutionary API

Figure 4a isolates the set of emulator's places encoding the P/T system's current state and structure. For example, as for the MS model, $M_0(\text{MARK}) = N \cdot \langle \text{In} \rangle$, where N is the number of pieces worked in a single production cycle. This set represents the emulator's *evolutionary interface*.

The adaptive behaviour of a system is described by a number of concurrent, user-defined procedures, which exploit a basic set of *read/write primitives*, called *evolutionary API*, to safely interface with the emulator. Each primitive acts on the emulator's evolutionary interface. The evolutionary API is a minimal but complete library for base-level *introspection/intercession*: one can get information about the marking and the graph structure of the P/T system, add/remove

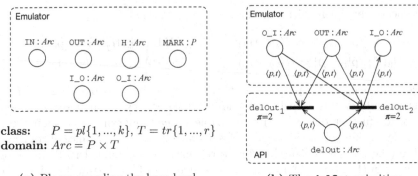

class: $P = pl\{1, ..., k\},\ T = tr\{1, ..., r\}$
domain: $Arc = P \times T$

(a) Places encoding the base-level.   (b) The delOut primitive.

Fig. 4. Emulator section and the delOut API primitive example.

P/T nodes, set the weight of arcs, change the current marking, etcetera. A primitive is defined by a SN subnet which reads and/or consistently modifies the P/T system's encoding in an atomic way, i.e., through a sequence of invisible actions. The evolutionary API is similar to the reflection API of most modern programming languages, as in the seminal idea introduced in [11] for (object) Petri nets.

Figure 4b shows a simple example of primitive (the delOut operation), which decreases the weight of a base-level output arc. When a token $\langle pr, tr \rangle$ is put into place delOut (holding the input of the primitive), one of the two mutually exclusive transitions delOut$_1$, delOut$_2$ may become enabled. Its firing removes the token $\langle pr, tr \rangle$ from the OUT place and updates the marking of I_O and O_I accordingly. The priorities of transitions composing a primitive sub-net are relative: when bringing all together, the greatest priority in a primitive-net is set lower than the lowest priority in the emulator.

Other primitives are more complex, due to additional consistency checks they perform. As an example, the primitive which decreases the weight of an input arc (the argument) has to check whether the linked transition is currently either in enabList or in checkList; if not, it has to be added to checkList.

4.3 Self-Adaptation Procedures

The managing subsystem, i.e., the part of the whole model driving the system evolution, is made up of a collection of adaptation procedures, each implementing a *feedback control loop* which deals with an adaptation concern. A procedure is described by a sub-net, which may contain both observable and logical transitions, and is indirectly connected to the emulator by means of the evolutionary API's input/output places. Figure 5 shows the procedure which, in the running example, manages a fault occurrence. This procedure refers to a simplified version of the MS running example, where just one line is faulty and it is periodically subject to failures.

This procedure is triggered whenever place Broken of the MS model is marked. A challenging point is that the MS execution keeps going while changes

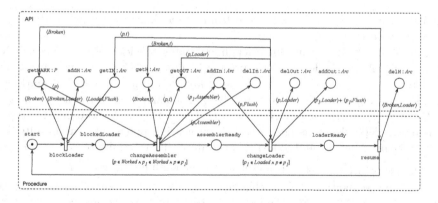

Fig. 5. The *fault managing* procedure.

are being carried out. Even though single transformations are atomic, the sequence used to apply them may bring the overall system into inconsistent states and thus affect the functional correctness.

A fault occurrence is checked by the blockLoader transition through the getMARK primitive. When blockLoader fires, it temporarily suspends the loader by linking it to place Broken with an inhibitor arc (addH primitive). Then changeAssembler modifies the arcs surrounding assembler, as shown in Fig. 2, through the addIN and delIN primitives. In a similar way, the procedure changes the loader behaviour to avoid loading of raw pieces into the faulty line (through addOut and delOut primitives). A new transition is inserted through which resid-ual row pieces on the faulty line eventually move to the working one. Loading is resumed at the end of the procedure by removing the temporary inhibitor arc between Broken and Loader. The procedure which brings the system back to its default layout (after the faulty line has been repaired) is not described due to lack of space. The tricky point there is that the system must enter a safe state before the reconfiguration can take place. The specification of this procedure is available on the online repository mentioned before.

The emulator, the evolutionary API, and the adaptation procedures are connected using a simple place *superposition*. The composition process can be automatically performed using the ALGEBRA package of GREATSPN.

5   Performance Analysis

The emulator-based model can be used for performance analysis if the stochastic extensions of PNs (SPNs), and symmetric nets (SSN) are used. Time specification and complexity issues have to be addressed. In SPNs, each transition t is associated with a *rate* $\rho(t) \in \mathbb{R}^+$ characterizing a non-negative exponential probability density function. A SPN is isomorphic to a Continuous Time Markov Chain (CTMC) whose states are the SPN reachable markings: the entry $[q_{i,j}]$ $(i \neq j)$ of the generator matrix is $\sum_{t:\, m_i[t > m_j} \rho(t)$ $([q_{i,i}] = -\sum_{i,j\, i \neq j} q_{i,j})$.

Performance indices can be computed from either the transient or steady state probability vector.

The ability to define marking-dependent rates considerably enhances SPN expressivity. A rate is defined by a product $\phi(t, m) = \rho(t) \cdot e(m)$, where $\rho(t)$ is the base rate of t, and $e(m) \in \mathbb{R}^+$ is the evaluation of a function e on marking m. The marking expression e may involve particular places of the net (usually, of ${}^\bullet t$), e.g., $p_1 + \frac{1}{2}p_2$ (hence $e(m) = m(p_1) + \frac{1}{2}m(p_2)$). A more generic type of marking dependency assigns a transition a firing policy, like the infinite/k-server, the mass-action, and so forth. For example, if t is infinite-server, $\phi(t, m) = \rho(t) \cdot min_{p \in {}^\bullet t} \left\lfloor \frac{m(p)}{I(p,t)} \right\rfloor$, the 2nd factor being the enabling degree of t.

The time semantics of a stochastic SN (SSN) [8] is that of its unfolding, a Generalized SPN (GSPN). Timed (i.e., observable) transitions are associated with exponential firing rates, as in SPN, whereas immediate (i.e., logical/black) transitions fire in zero-time. The latter are assigned weights to probabilistically determine which transition fires, in the case of simultaneous enabling. In SSN, rate/weights may be associated with transition instances. A function ω defines the rates/weights of a timed/immediate transition, as follows ($c \in cd(t)$):

$$\omega(t, c) = \begin{cases} r_i & if\ cond_i(c),\ i = 1, \ldots, n; \\ r_{n+1} & otherwise \end{cases}$$

where $cond_i$ is a boolean expression built of standard predicates on the transition's color instance. Hence, the firing rate/weight $r_i \in \mathbb{R}^+$ of a transition instance can only depend on the static subclasses of the colors assigned to the transition variables and on the comparison of variables of the same type. We assume that the conditions $cond_i$ are mutually exclusive.

The stochastic process underlying a SSN model is a CTMC, whose states are identified with the SSN *tangible* markings. The entry $[q_{i,j}]$ ($i \neq j$) of the generator matrix is $\sum_{t \in T, c \in cd(t), \sigma: M_i[(t,c) \cdot \sigma > M_j} \omega(t, c)\mathbb{P}\sigma$, where σ is a (possibly null) immediate transition sequence and $\mathbb{P}\sigma$ the corresponding probability (1 if σ is empty).

Marking dependent rates/weights may be defined, according to the static partitioning of basic color classes, through a function $\varphi(M, t, c) = \omega(t, c) \cdot \hat{e}(M)$, where the 2nd element of the product is the evaluation of a (symbolic) marking expression in which static subclasses are used instead of colors. Assuming, e.g., $cd(p_1) = C$, $cd(p_2) = C \times C$, $C = C_1 \cup C_2$: the expression $\hat{e} = \frac{p_2(\langle C_1, C_2 \rangle)}{max(1, p_1(C_1))}$, when evaluated on M, gives the ratio between the number of tokens in place p_2 with 1st element of subclass C_1 and 2nd of subclass C_2, and the max between 1 and the number of tokens of subclass C_1 in place p_2[2].

5.1 Emulator's Colour Class Partitioning

We have seen that the untimed behaviour of a P/T system is exactly reproduced by the emulator encoding it. In order for the emulator to preserve also the time

---

[2] (in)equalities between colors of the same subclass may be expressed through $cond_i$.

semantics of the encoded SPN, the emulator's color classes may have to be partitioned. In fact, a SPN transition tr corresponds to the binding $(t = tr)$ of SN *timed* transition PT_fire (Fig. 3).

No Marking Dependency or Use of Generic Firing Policies in the SPN. In these cases, we may only have to partition color class T into $\bigcup_i^n T_i$, such that $\forall i$: $T_i \supseteq \{tr_j \in T | \rho(tr_j) = r_i\}$ and $\omega(\text{PT\_fire}, t = tr_j) = r_i$ if $t \in T_i$.

A generic firing policy at SPN level is expressed in terms of emulator's places MARK and IN. As for the infinite-server, the enabling degree of tr becomes

$$\min_{pl:\, \text{IN}(\langle pl, tr \rangle) \neq 0} \left\lfloor \frac{\text{MARK}(p_l)\rfloor}{\text{IN}(\langle pl, tr \rangle)} \right\rfloor$$

Transitions in a given subclass must be characterized by the same firing policy.

Use of General Marking Dependency in the SPN–If we want to reflect this kind of dependency in the emulator, we may have to partition also color class P. Let e_{tr} be the expression defining the marking dependency of a SPN transition. This expression involves *some* SPN places. The idea is to partition classes T and P into $\bigcup_i T_i$ and $\bigcup_j P_j$, such that the condition above holds and, letting $\hat{e}_{tr}$ be the symbolic expression obtained from e_{tr} by replacing every place symbol with the static subclass it belongs to, $\forall i, tr \in T_i : \hat{e}_{tr} = \hat{e}_{T_i}$, where $\hat{e}_{T_i}$ is the transition subclass marking expression.

Marking Dependency Issues–The use of marking dependency in a dynamic context, however, has got some trickiness. The possibility of withdrawing places from the encoded net may lead to a situation in which, e.g., all the places of a given subclass disappear, whereas there is some transition which refers to it in the marking dependency pattern. Emptying the pre-set of a SPN transition with an infinite- or k-server semantics is another cause of incongruence. To avoid such situations, we have enriched the primitives of the evolutionary API removing places/input arcs with simple additional controls based on subclasses.

Assignment of Weights to Immediate Transitions–Due to property 2 at the end of Sect. 4 (independently of the order in which immediate transitions fire, from a given TM we always reach the same TM), this has no relevance on the timed behaviour of the emulator. Weights may therefore be arbitrarily assigned. Summarizing (possibly after a partitioning of the emulator's color classes), the following claim holds: any transition from a state m to m' of the encoded SPN characterized by rate λ is matched by a transition between corresponding tangible states M, M' of the emulator with the same rate.

The weights assigned to the immediate transitions of the evolutionary API's primitives do not influence the model's stochastic behaviour because of the absence of conflicts among them.

6   Experiments

Table 1 reports some experiments conduced on the running example, the self-healing MS, by using GreatSPN. The whole model, composed of the emulator

Table 1. Reachability graph size/building time and transition throughputs.

Model	N	$|RG|$ (TM/VM)	Time (s.)	$|SRG|$ (TM/VM)	Time (s.)	Throughput	Time (s.)
MS	2	55/3484	0.15	55/3484	3.40	0.20074	0.66
	4	184/13906	2.31	184/13906	10.06	0.26590	1.12
	8	985/91586	8.04	985/91586	64.11	0.33175	5.36
	16	7964/886842	77.11	7964/886842	622.91	0.38866	49.71
	32	95568/11108025	1544	95568/11108025	≈3.5 h	0.43867	560.11
SMS	2	92/6708	1.71	48/3437	5.36	0.19532	0.93
	4	276/23268	2.79	142/11986	10.06	0.25852	1.88
	8	1289/131761	10.12	662/66663	91.66	0.32164	5.59
	16	9103/1114027	99.05	4634/561461	816.71	0.37521	62.27
	32	109236/13945378	1978	55448/7078872	≈4.8 h	0.41121	720.30

(initially encoding the P/T system in Fig. 1) and the two adaptation procedures, has been analysed for values of N (number of worked pieces per cycle) 2 to 32. Since the model's TRG have a cyclic structure (their initial markings are home states), a steady-state solution of the corresponding CTMS does exist. Transition firing rates (irrelevant, for our scopes) are those indicated in Figs. 1 and 2. All transitions are assumed to be infinite-server. The color class T, which holds as many colours as the union of transition in Figs. 1 and 2, has been partitioned into subclasses characterized by the same firing rate, as described in Sect. 5. Given that GreatSPN doesn't currently support a color-dependent definition of rates, timed transition PT_fire has been partially unfolded into a set of mutually exclusive instances, each associated with a guard $t \in T_i$, where T_i is a subclass (e.g., $\{line_1, line_2\}$). The *throughput* column shows the average throughput of the *Assembler* transition, that measures the system efficiency and (when compared to the nominal MS behaviour in absence of faults) the overhead due to reconfiguration. Two variants of the MS model are considered: the symmetric one (denoted SMS) described in Fig. 1, and the asymmetric one (MS), in which only one of the two lines may be periodically off. As discussed later, some data refer to the solution of a lumped CTMC directly derived from (the tangible part of) a quotient graph of the ordinary RG, called Symbolic Reachability Graph (SRG), which exploits the (possible) model symmetries. The throughput values are congruently the same for the ordinary CTMC and the lumped one. The last column of Table 1 reports the time required to compute the throughput. This value ranges from a few milliseconds to a few hundred seconds. The size of the RG and SRG are listed in terms of tangible and vanishing states. Execution times are reported, varying from dozens ms to a few hours. We observe that, as discussed later, the SRG takes much more time than the RG.

7   Facing Complexity

The high number of immediate transitions in the SN emulator may leads to an explosion of vanishing markings, as evident from Table 1, thus affecting the model's solution. The state-space builder implemented in GREATSPN has, in fact, some drawbacks. In particular, it does not use any on-the-fly reduction of immediate transition paths, i.e., the vanishing markings are eliminated (to get the corresponding TRG, and the associated CTMC) after the whole reachability set of markings is built. Moreover, simultaneously enabled immediate SN transitions are fired in an interleaved way (causing a combinatorial explosion of vanishing states), even when their instances are independent. The first limitation might be faced only reimplementing the state-space builder. The inefficiency caused by the interleaving of immediate transitions instances has been tackled by using two orthogonal approaches, shortly discussed in the following.

Ordering/Partitioning of Basic Classes–The main source of inefficiency is the way in which the enabling test of P/T transitions is performed (*step i*), that is, through an interleaving of the instances of transition nextT, which is in charge of selecting the next P/T transition to be checked for. By the way, the order in which these transitions are considered is irrelevant, therefore the interleaving can be significantly reduced by defining the color class T as *ordered*. This solution, the one used in the emulator in Fig. 3, drops the number of immediate firing sequences (potentially) from $n!$ to n, where n is the cardinality of class T. Just to give an idea, for $N = 32$ the number of vanishing markings (SMS model) lowers to around one million, and the RG building time to 300 s. We might analogously set the class P as ordered, so to consider input/inhibitor/output places in an arbitrary order during both the enabling and firing steps, even if the achievements should be less evident, but for particular cases. Although ordering colour classes is effective, it unfortunately prevents from exploiting symmetries in performance analysis, except in the extremely rare circumstance in which no static partitioning of classes is required: in fact, ordering a partitioned class implicitly causes its complete splitting into singleton subclasses.

An alternative solution is to exploit the partition of color class T into subclasses, induced by the SPN model's time specification, to reduce the interleaving of nextT. This simple idea is described in Fig. 6 (for a generic case), showing the portion of the SN emulator that has to change accordingly: transition nextT has been split into a number of mutually exclusive instances, each one associated with a guard testing the membership of a P/T transition to a specific subclass. Transitions belonging to different subclasses are considered in an arbitrary order, by assigning the partially unfolded instances of nextT different priorities. The gain, in terms of interleaving reduction, depends on the size of the biggest subclass (the smaller, the better). As for the (S)MS example, where this size is two, the achieved reduction of vanishing states is slightly lower than the one achieved with the ordering of class T.

Structural Techniques–[7] recently introduced a calculus (and a working implementation[3]) for deriving *symbolic structural relations* between SN nodes, in particular SC, and CC, that may help build efficiently the reachability graph. Structural relations between SN transitions are defined as mappings $cd(t) \rightarrow 2^{cd(t')}$. For example $SC(t, t')$ (the asymmetric Structural Conflict) maps an instance

---

[3] Available at http://www.di.unito.it/~depierro/SNex/.

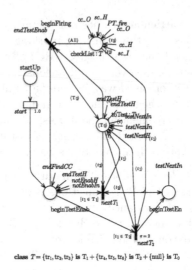

class $T = \{tr_1, tr_2, tr_3\}$ **is** $T_1 + \{tr_4, tr_5, tr_6\}$ **is** $T_2 + \{null\}$ **is** T_0

Fig. 6. Split of nextT due to the partition of T into $T_1 \cup T_2$.

(t, b) to the set of instances $\{(t', b')\}$ that can disable (t, b) by withdrawing color-tuples from some input place of t or adding color-tuples into some inhibitor place of t. Such relations are *syntactically* expressed by using a simple extension of SN arc functions' grammar. By computing SC, CC, and the transitive closure, it is possible to check whether two transitions t and t' are *structurally independent*, meaning that there is no instance of t conflicting (either directly, or indirectly through a sequence of causally connected higher priority transitions) with any instance of t', and vice versa. Independent SN transitions have been assigned different priorities, to reduce interleaving. It is worth noting that it is also possible to decrease the interleaving of instances of the same transition, which is a major concern in the SN emulator. For instance, since $SC(\text{testNextIn}, \text{testNextIn}) = \emptyset$, the instances of testNextIn (which selects a place from a P/T transition's preset) might be fired in any order. Structural analysis has been also used to validate the emulator-based model. In particular, we exploited it to prove the absence of conflicts among immediate transition instances of the evolutionary API subnets, what makes the assignment of weights to this component irrelevant from the performance analysis point of view.

Symmetry Exploitation–By setting an initial *symbolic* marking it is possible to build a *quotient-graph*, called symbolic reachability graph (SRG), which retains all the information of the ordinary RG. SRG nodes are *syntactical* equivalence classes of ordinary colored markings, where m, m' are equivalent if and only if m' is obtained from m through a permutation on basic classes preserving the partition into subclasses and the circular ordering. In stochastic SN a CTMC is derived from the SRG, whose states denote aggregates for which both the *exact* and *strong* lumpability hold. This reduced CTMC can be solved, instead of the original one. A symbolic marking (SM) is defined in terms of *dynamic subclasses*.

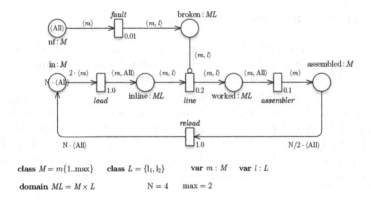

Fig. 7. A symmetric MS model

Each dynamic subclass refers to a static subclass, or to a basic class (in the case of a non partitioned class), and has a cardinality. Dynamic subclasses represent *parametric partitions* of color (sub-)classes. Dynamic subclasses of an ordered class are ordered too. A simple way to set up an initial symbolic marking in the emulator is to replace in the ordinary initial marking colors $\{pl_i\}$ and $\{tr_i\}$ with (cardinality one) dynamic subclasses $\{zp_i\}$ and $\{ztr_i\}$, respectively. The resulting SRG nodes (SMs) represent classes of isomorphic *marked P/T nets*. Checking graph isomorphism is a demanding task and in our model corresponds to bring an SM (which encodes a graph) into its canonical form [8]. Table 1 shows that in the case of symmetric MS the SRG size, as expected, is more or less the half of the ordinary RG. On the other side, there is an evidence that building the SRG is much more time consuming than building the ordinary RG. We believe that a major source of inefficiency, in the current implementation, is that the cardinality of an SM has to be computed, and this is done by explicitly enumerating the possible permutations represented by the SM. In order to alleviate this problem, it is convenient to further refine the partitions of classes T and P induced by time specification, so that each subclass contains nodes which are known a priori as permutable.

A possible way to automatically derive this information is starting from a symmetric SPN, like that described in Fig. 7, which represents a parametric version of the MS system with *max* copies of the MS. The idea is that all and only the P/T nodes which are the unfolded instances of a SSN node would be gathered in the same subclass.

Analysing models composed of a large number of identical modules, on the other hand, is unfeasible without exploiting symmetries. Think, e.g, that a configuration with 4 MS components, each working 4 pieces per cycle, results in several dozens millions states, against just a few thousands symbolic ones.

8   Conclusion and Future Work

We have introduced a SSN-based model able to emulate SPNs with changing layout. The approach exploits SN analysis capabilities and is supported by off-the-shelf analysis tools like GREATSPN. A self-healing MS has been used as a running example to point out benefits/drawbacks of the model. Major complexity issues have been faced by using two complementary techniques based on ordering/partitioning of basic color classes and computation of symbolic structural relations, respectively. We plan to fully automate the modelling process and develop an optimized state-space builder by leveraging on-the-fly reduction of vanishing markings and structural techniques.

References

1. Baarir, S., Beccuti, M., Cerotti, D., De Pierro, M., Donatelli, S., Franceschinis, G.: The GreatSPN tool: recent enhancements. SIGMETRICS Perform. Eval. Rev. **36**(4), 4–9 (2009). https://doi.org/10.1145/1530873.1530876
2. Cabac, L., Duvigneau, M., Moldt, D., Rölke, H.: Modeling dynamic architectures using nets-within-nets. In: Ciardo, G., Darondeau, P. (eds.) ICATPN 2005. LNCS, vol. 3536, pp. 148–167. Springer, Heidelberg (2005). https://doi.org/10.1007/11494744_10
3. Camilli, M., Gargantini, A., Scandurra, P.: Specifying and verifying real-time self-adaptive systems. In: 2015 IEEE 26th International Symposium on Software Reliability Engineering (ISSRE), pp. 303–313, November 2015. https://doi.org/10.1109/ISSRE.2015.7381823
4. Camilli, M., Capra, L., Bellettini, C.: PNemu: an extensible modeling library for adaptable distributed systems. In: Donatelli, S., Haar, S. (eds.) PETRI NETS 2019. LNCS, vol. 11522, pp. 80–90. Springer, Cham (2019). https://doi.org/10.1007/978-3-030-21571-2_5
5. Capra, L.: A pure SPEC-inscribed PN model for reconfigurable systems. In: 2016 13th International Workshop on Discrete Event Systems (WODES), pp. 459–465, May 2016. https://doi.org/10.1109/WODES.2016.7497888
6. Capra, L., Camilli, M.: Towards evolving petri nets: a symmetric nets-based framework. IFAC-PapersOnLine **51**(7), 480–485 (2018). https://doi.org/10.1016/j.ifacol.2018.06.343. 14th IFAC Workshop on Discrete Event Systems WODES 2018
7. Capra, L., De Pierro, M., Franceschinis, G.: Computing structural properties of symmetric nets. In: Campos, J., Haverkort, B.R. (eds.) QEST 2015. LNCS, vol. 9259, pp. 125–140. Springer, Cham (2015). https://doi.org/10.1007/978-3-319-22264-6_9
8. Chiola, G., Dutheillet, C., Franceschinis, G., Haddad, S.: Stochastic well-formed coloured nets for symmetric modelling applications. IEEE Trans. Comput. **42**(11), 1343–1360 (1993)
9. Hoffmann, K., Ehrig, H., Mossakowski, T.: High-level nets with nets and rules as tokens. In: Ciardo, G., Darondeau, P. (eds.) ICATPN 2005. LNCS, vol. 3536, pp. 268–288. Springer, Heidelberg (2005). https://doi.org/10.1007/11494744_16
10. Jensen, K., Rozenberg, G. (eds.): High-level Petri Nets: Theory and Application. Springer, London (1991). https://doi.org/10.1007/978-3-642-84524-6
11. Lakos, C.: Towards a reflective implementation of object petri nets. In: Proceedings of TOOLS Pacific, pp. 129–140 (1996)

12. Lakos, C.: Object oriented modelling with object petri nets. In: Agha, G.A., De Cindio, F., Rozenberg, G. (eds.) Concurrent Object-Oriented Programming and Petri Nets. LNCS, vol. 2001, pp. 1–37. Springer, Heidelberg (2001). https://doi.org/10.1007/3-540-45397-0_1

13. Reisig, W.: Petri Nets: An Introduction. Springer, New York (1985). https://doi.org/10.1007/978-3-642-69968-9

14. Tigane, S., Kahloul, L., Benharzallah, S., Baarir, S., Bourekkache, S.: Reconfigurable GSPNs: a modeling formalism of evolvable discrete-event systems. Sci. Comput. Program, 102302 (2019). https://doi.org/10.1016/j.scico.2019.102302

15. Valk, R.: Object petri nets. In: Desel, J., Reisig, W., Rozenberg, G. (eds.) ACPN 2003. LNCS, vol. 3098, pp. 819–848. Springer, Heidelberg (2004). https://doi.org/10.1007/978-3-540-27755-2_23

16. Weyns, D., Iftikhar, M.U., de la Iglesia, D.G., Ahmad, T.: A survey of formal methods in self-adaptive systems. In: Proceedings of the Fifth International C* Conference on Computer Science and Software Engineering, C3S2E 2012, pp. 67–79. ACM, New York (2012). https://doi.org/10.1145/2347583.2347592

Design and Evaluation of an Edge Concurrency Control Protocol for Distributed Graph Databases

Paul Ezhilchelvan[1]([✉]), Isi Mitrani[1]([✉]), Jack Waudby[1]([✉]), and Jim Webber[2]([✉])

[1] School of Computing, Newcastle University, Newcastle upon Tyne NE4 5TG, UK
{paul.ezhilchelvan,isi.mitrani,j.waudby2}@ncl.ac.uk
[2] Neo4j UK, Union House, 182-194 Union Street, London SE1 0LH, UK
jim.webber@neo4j.com

Abstract. A new concurrency control protocol for distributed graph databases is described. It avoids the introduction of certain types of inconsistencies by aborting vulnerable transactions. An approximate model that allows the computation of performance measures, including the fraction of aborted transactions, is developed. The accuracy of the approximations is assessed by comparing them with simulations, for a variety of parameter settings.

Keywords: Graph databases · Reciprocal consistency · Edge-order consistency · Arbitration · Stochastic modelling · Simulation

1 Introduction

Existing large-scale distributed data stores such as Google Docs, Dynamo [4] and Cassandra [3] implement an 'eventually consistent' update policy (see [14]). That is, update requests are processed as soon as they arrive. In some cases this is a reasonable choice. For a non-partitioned system there are several solutions to dealing with what is effectively lag between replicas. However, when a database is partitioned among several hosts, the eventual consistency approach raises serious problems, especially when there are explicit or (application) implied relationships between the data stored in different partitions.

For example, a patient might observe an appointment has been booked in their timeline on partition A, while the corresponding clinician in partition B hasn't yet blocked off that slot. Eventual consistency makes it possible for another patient to book into that slot either overwriting or double-booking the clinician. While each partition on its own will be eventually consistent, the system as a whole has violated a constraint.

This is similar in a sense to problems that can occur in traditional databases with Snapshot Isolation (SI), but unlike SI there are no mechanisms in eventually consistent databases to detect distributed constraint violations. For distributed graph databases this is a critical problem because explicit relationships (edges)

M. Gribaudo et al. (Eds.): EPEW 2019, LNCS 12039, pp. 50–64, 2020.
https://doi.org/10.1007/978-3-030-44411-2_4

routinely span across partitions, and unless both partitions agree reciprocally on the existence, direction, and content of the edge then the database has become corrupted.

Another example, by Bailis and Ghodsi [2] refers to an ATM service where eventual consistency can allow two users to simultaneously withdraw more money than their (joint) bank account holds; such an anomaly, on being detected, is reconciled by invoking exception handlers. Given that an ATM service is expected to be available 24/7 and that account holders are permitted to access only their own accounts, the eventually consistent approach is appropriate.

A vast majority of common graph database applications, however, allow data modified by one (user) transaction to be read by an arbitrary number of other (users') transactions (see Robinson et al. [12]). In such cases, data corrupted by one transaction and read by subsequent transactions, can lead to further corruption from which it is impossible to recover. This process, which was studied at some detail by Ezhilchelvan et al. [5], can in time cause the entire database to become unusable. That is a situation that is certainly worth avoiding.

In this paper we propose a new update protocol where conflicting updates are detected and handled. Corruption is thus prevented, but the price paid for this improvement is that some transactions are aborted. In order to evaluate just how heavy is that price, we also construct and analyse an approximate model that allows us to compute the average number of transactions aborted per unit time and other performance measures.

To simplify the protocol presentation and analysis, we assume that the hosts are reliable and data items in a database are not replicated. Provisions for crash-tolerance can be incorporated as an orthogonal aspect by using well-known techniques (e.g., single server abstraction) and supportive technologies (e.g., Raft [8], Paxos [7]).

The problem context and the proposed protocol are described in Sects. 2 and 3. The approximate model and its analysis are presented in Sect. 4. Some numerical and simulation results are reported in Sect. 5, while Sect. 6 outlines the conclusions.

2 Problem Description

A graph database consists of *nodes* representing entities, and *edges* representing relations between them (see [12]). For example, node X may represent an entity of type Author and Y an entity of type Book. X and Y will have an edge between them if they have a relation, e.g. X is an author of Y.

The popularity of the graph database technology owes much to this simple structure from which sophisticated models can be easily built and be efficiently used for query or transaction processing. Examples of operations performed on a graph database are: finding shortest paths between two locations in a transport network, performing product recommendations, looking for cancerous patterns in biological data, etc.

When nodes are connected by an edge, the database stores some reciprocal information at the origin and destination records of that edge. For example, if there is an edge from node X to node Y, then node X would have an *outgoing* record *wrote* and node Y would have an *incoming* record *wrote*, which can be interpreted as *written by*. Maintaining this reciprocal information enables an edge to be traversed in either direction.

An edge e is said to be *reciprocally consistent*, if its origin and destination records, denoted as e_1 and e_2, at the nodes that it connects, have mutually consistent, reciprocal entries.

In a distributed graph database, graph data is partitioned and each partition is hosted by a server in a cluster. Partitioning a graph is non-trivial and even the most optimal partitioning algorithms (e.g., [10,11]) seek only to minimise, and cannot eliminate, the presence of distributed edges. The outgoing and incoming records of a distributed edge are on different hosts[1]. It has been estimated in [13], that a fraction varying between 25% and 75% of all edges would be distributed. Maintaining reciprocal consistency across a distributed edge is challenging because its e_1 and e_2 records cannot be updated simultaneously. The time interval that elapses between those updates permits interference among concurrent transactions.

Suppose, for example, that nodes F and S, referring to a flight and a seat in an airline database, are stored on hosts $H1$ and $H2$ respectively, with the edge between them indicating availability. Two transactions, U and V, write 'S is available in F' and 'S is booked in F', respectively. Each update operation is carried out first on one of the hosts and then, after a small but non-zero 'network delay', on the other host. These two phases of the update are referred to as 'part 1' and 'part 2', respectively. The delay interval between them, D, is a random variable which may, in principle, be unbounded.

Such an implementation, if left uncontrolled, makes possible the introduction of faults in the edge records. This is illustrated in Fig. 1, which shows three possible conflict scenarios between transactions U and V (time flows downwards). In case (a), transaction U performs part 1 of the update on $H1$ at time t and part 2 on H_2 at time $t + D$. At some point between t and $t + D$, transaction V performs part 1 on H_2, and part 2 on H_1 some time later. The result of this occurrence is a violation of reciprocal consistency: the H_1 entry ends up saying 'seat S is booked in F', while the H_2 entry says 'seat S is available in F'.

A similar conflict is shown in case (b), except that here part 1 of V is performed on H_2 *before* time t, and part 2 on H_1 after t. Finally, there is the possibility (c), where both transaction traverse the edge in the same direction, but U overtakes V during the network delay. The result of that conflict is that $H1$ claims 'seat S is available in F', while H_2 says 'seat S is booked in F'.

In order to prevent such conflicts, only transactions whose distributed updates are 'interference-free', should be allowed to proceed. That is, if part 1 of an update for a given edge precedes part 1 of another update for the same edge, then so should part 2, and vice versa. One could, of course, avoid such conflicts

---

[1] This is avoided in *edge-partitioned* graph databases, where all instances of a given edge type reside in the same partition, and nodes are replicated. Then the problem is to ensure that updates to nodes are consistent across partitions.

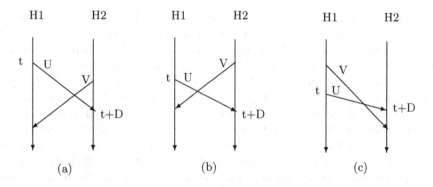

Fig. 1. Three possible conflict scenarios

by using a strong consistency mechanism, but the availability and throughput penalties are generally prohibitive.

A different type of possible conflicts arises when transactions update more than one edge during their lifetime. For example, suppose that transactions A and B both update edges e and e', and do so without interference either among themselves or with other transactions. It may happen that e is updated by A before B, while e' is updated by B before A, as illustrated in Fig. 2 (here time flows from left to right and the conflict-free updates are collapsed to single instants). Such an occurrence, if allowed, would violate the property of 'edge-order consistency' between transactions.

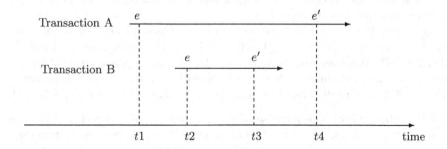

Fig. 2. Edge-order consistency violation

Inconsistencies of this type are to be avoided because they compromise an important database property known as serializability.

3   Edge Concurrency Control Protocol

Our protocol employs two distinct mechanisms, referred to as 'collision detection' and 'order arbitration', respectively. These are aimed at enforcing

(i) reciprocal consistency for distributed updates and (ii) edge-order consistency between transactions. Collision detection is applied at every update and may trigger an immediate abort. Transactions that survive collision detection may, if necessary, go to order arbitration. The latter may also result in an abort.

Like many concurrency control protocols in the literature (e.g., [6,9]), our protocol treats updates as being provisional initially. They become permanent only if, and when, the transaction that contains them is allowed to commit.

Provisional updates on a given record can occur only one at a time and each is time-stamped using the local host's clock. Thus, when a transaction attempts to update a given record, it can identify all other transactions, called *predecessors* (if any), that have earlier updated that record provisionally. Read operations receive the latest committed version of a record and ignore any provisionally updated values.

Collision Detection. Remember that an update operation by a transaction for a distributed edge has a part 1, carried out at the first host visited, and part 2, performed at the second host after a network delay. The corresponding provisionally updated records are now given labels 1 and 2, respectively, and are associated with the id of that transaction. These labels act as 'history' meta-data indicating the host where a transaction started and completed its update. The following rule is applied:

Cancellation Rule: If, by the time part 2 is performed, a previous provisional update labeled 1 has been observed, but the corresponding label 2 has not been observed, then this update is cancelled. In other words, an update is cancelled if it has observed the start of a previous attempt, but not its completion.

When an update is cancelled, the transaction containing it is aborted and all its provisional records are erased.

According to the this rule, update U in Fig. 1, cases (a) and (b), is cancelled because it observes a predecessor V with label 1 in $H2$, but has not observed V's label 2 in $H1$. Whether update V is cancelled or not, depends on whether U's provisional updates remain or have been erased by the time V performs part 2. In case (c), U is cancelled but V is not, because it does not observe U's label 1.

Order Arbitration. The purpose of this mechanism is to detect and prevent edge-order inconsistencies between transactions. It only applies to transactions that contain more than one distributed update. Those with a single update that have not been aborted by collision detection are allowed to commit and depart.

Using the records relating to provisional updates, each multi-update transaction maintains a 'predecessor list' containing all predecessor transactions it encountered during its provisional updates. If that list is empty when the transaction successfully completes all its provisional updates, then it commits and departs. Otherwise it goes to arbitration.

The arbiter is a special service, assumed here to have been implemented in a dedicated host. A list called the *hit-list* is maintained. It contains transactions which, if allowed to commit, risk violating edge-order consistency. Transactions arriving for arbitration join a queue and are served in order of arrival. If the transaction at the head of the queue is not present in the hit-list, it commits. All transactions in its predecessor list are added to the hit list if not already there. If it is in the hit list, it aborts and all its provisional updates are erased. What has happened in this case is an overtaking: the current transaction was named as a predecessor by a transaction that committed earlier.

We can informally argue that our approach is correct by considering the edge-order inconsistency depicted in Fig. 2. It can be seen that A will have B in its predecessor list while updating e', and B will observe A as a predecessor while updating e. Both A and B must approach the arbiter because they update more than one edge and have a non-empty predecessor list. If the first transaction to be processed by the arbiter is allowed to commit, the second one will be entered in the hit list and will abort. Thus only one of A and B, but not both, can commit and edge order inconsistency is always avoided.

Note that this approach to arbitration is pessimistic. It aborts a transaction as soon as it detects a risk of edge-order violation, even though the actual violation may not occur. Consequently, some transactions are aborted unnecessarily, just because they are overtaken by their successors. To eliminate unwarranted aborts, the arbiter would have to keep much more detailed information about the updates performed by all transactions, and would have to do considerably more processing.

We now proceed to the task of evaluating certain performance measures, such as the average number of transactions that are aborted per unit time, the offered load at the arbiter, and the average time a transaction remains in the system. Since the processes involved are rather complex, such an evaluation will inevitably entail approximations. That, in turn, will necessitate an assessment of the accuracy of those approximations.

4   Approximate Model

We are concerned with updates performed on distributed edges in a graph database (i.e., edges whose source and destination nodes are stored on different hosts). These edges are divided into T types, numbered 1, 2, ..., T. The number of edges of type i is N_i, and the probability that an update operation is aimed at an edge of type i is p_i. All edges of a given type are equally likely to be addressed, so that the probability of accessing a particular edge of type i is p_i/N_i.

Transactions arrive into the system in a Poisson stream, at the rate of λ per second. Each transaction performs a random number, K, of updates for different distributed edges. The distribution of K is arbitrary: $P(K = k) = r_k$ $(k = 1, 2, \dots)$. The average number of updates per transaction is κ. Thus, the arrival rate of updates at a *particular* distributed edge of type i, ξ_i, is equal to

$$\xi_i = \frac{\kappa \lambda p_i}{N_i}; \ i = 1, 2, \dots, T. \tag{1}$$

The first approximation is to assume that the arrival process of updates for a particular edge of type i is Poisson with rate ξ_i.

We wish to estimate the probability, u_i, that an update, U, for an edge of type i, is cancelled due to a collision with another update, V, for the same edge. That is, either V arrives in the opposite host during the network delay of U (Fig. 1, case (a)), or U arrives in the opposite host during the network delay of V (case (b)), or U arrives in the same host during the network delay of V *and* its network delay completes before that of V (case (c)).

Assume that the network delays are i.i.d random variables distributed exponentially with parameter δ (mean $1/\delta$). This may or may not be an approximation.

Updates for a particular edge of type i arrive in a particular one of the two hosts involved at rate $\xi_i/2$. Moreover, a given network delay completes before another with probability $1/2$. Hence, we can estimate the probabilities of cases (a), (b), and (c), $u_i^{(a)}$, $u_i^{(b)}$ and $u_i^{(c)}$, as

$$u_i^{(a)} = u_i^{(b)} = \frac{\xi_i}{\xi_i + 2\delta}; \quad u_i^{(c)} = \frac{1}{2}\frac{\xi_i}{\xi_i + 2\delta}; \quad i = 1, 2, \ldots, T, \tag{2}$$

where ξ_i is given by (1) and δ is the parameter of the network delay. The overall probability, u_i, that at least one of those events will happen, is

$$u_i = 1 - (1 - u_i^{(a)})(1 - u_i^{(b)})(1 - u_i^{(c)}) \approx \frac{2.5\xi_i}{\xi_i + 2\delta}; \quad i = 1, 2, \ldots, T. \tag{3}$$

The last approximation in the right-hand side holds when the rate ξ_i is small compared to δ.

The unconditional probability, u, that an arbitrary update is cancelled by the collision detection mechanism, is given by

$$u = \sum_{i=1}^{T} p_i u_i. \tag{4}$$

The probability, v_k, that a transaction containing k updates is aborted because one of them is involved in a collision, is equal to

$$v_k = 1 - (1 - u)^k, \tag{5}$$

and the unconditional probability, v, that a transaction is aborted due to a collision is given by

$$v = \sum_{k=1}^{\infty} r_k v_k. \tag{6}$$

Now consider the average run time, a_k, of a transaction that contains k update operations. Assume that each update takes time b, on the average. Those times include read operations and computations, as well as network delays. If the first $j - 1$ provisional updates are completed successfully but the j-th update is cancelled as a result of a collision, then the average run time would be jb. Hence, a_k is given by

$$a_k = \sum_{j=1}^{k} jb(1 - u)^{j-1}u + kb(1 - u)^k, \tag{7}$$

where u is given by (4)

With a little manipulation, this expression can be simplified to

$$a_k = b \sum_{j=1}^{k} (1-u)^{j-1} = b\frac{1-(1-u)^k}{u}. \tag{8}$$

The unconditional average run time of a transaction, a, is equal to

$$a = \sum_{k=1}^{\infty} r_k a_k. \tag{9}$$

If all provisional updates in a transaction are completed successfully, and if either there was only one update, or there were no predecessors, then the transaction commits. Otherwise it goes to the arbiter. The time that a transaction spends queueing and being served by the arbiter will be referred to as the 'arbitration time'.

Assume (this is another approximation) that each transaction joins the arbiter queue with probability α, independently of the others. That is, the arrival process is Poisson, with rate $\lambda\alpha$. The arbiter's average service time, s, is a given parameter. Thus the offered load at the arbiter is $\rho = \lambda\alpha s$.

Treating the arbiter as an $M/M/1$ queue, we estimate the average arbitration time, w, as

$$w = \frac{s}{1-\rho}, \tag{10}$$

provided that $\rho < 1$. If $\rho \geq 1$, then $w = \infty$. The total average time that a transaction spends in the system is

$$W = a + \alpha w, \tag{11}$$

where a is given by (9).

We shall now develop an iterative fixed-point approximation for α. Denote by $d_{j,k}$ the average lifetime of the j'th update within a transaction containing k updates, *excluding* any possible arbitration time. By an argument similar to the one that led to (8), we obtain

$$d_{j,k} = b \sum_{i=1}^{k+1-j} (1-u)^{i-1} = b\frac{1-(1-u)^{k+1-j}}{u}. \tag{12}$$

The lifetime of a randomly chosen update within a transaction containing k updates, d_k (again excluding arbitration), is given by

$$d_k = \frac{1}{k} \sum_{j=1}^{k} d_{j,k} = b\frac{(k+1)u + (1-u)^{k+1} - 1}{ku^2}. \tag{13}$$

Hence, the total average time spent in the system by an arbitrary update (*including* the arbitration time), d, is equal to

$$d = \sum_{k=1}^{\infty} r_k d_k + \alpha w, \tag{14}$$

where w is given by (10).

Now, let γ_i be the probability that an update of type i has a predecessor, i.e. the probability that such an update arrives while a preceding update for the same edge is still in the system. Assuming that the update residence times are distributed exponentially with mean d given by (14), this can be approximated as

$$\gamma_i = \frac{\xi_i d}{1 + \xi_i d}, \tag{15}$$

where ξ_i is given by (1).

The unconditional probability, γ, that an arbitrary update has a predecessor, is

$$\gamma = \sum_{i=1}^{T} p_i \gamma_i. \tag{16}$$

If a transaction contains k updates, the probability that at least one of them has a predecessor, α_k, is

$$\alpha_k = 1 - (1 - \gamma)^k. \tag{17}$$

Remembering that a transaction goes to the arbiter if it has more than one update *and* all updates avoid collisions *and* at least one of them has a predecessor, we write

$$\alpha = \sum_{k=2}^{\infty} r_k (1 - u)^k \alpha_k. \tag{18}$$

Note that the right-hand side of (18) depends on α, via (14) and (15). In other words, we have a fixed-point equation of the form

$$\alpha = f(\alpha). \tag{19}$$

This can be solved by a simple iterative scheme. Start with an initial guess, α_0, say $\alpha_0 = 0$. At iteration n, compute

$$\alpha_n = f(\alpha_{n-1}), \tag{20}$$

stopping when two consecutive iterations are sufficiently close to each other.

The probability α allows us to evaluate the offered load at the arbiter queue, and hence estimate the average response time of a transaction, W. Another important performance measure is the rate of aborts, R, i.e. the average number of transactions that are aborted per unit time. Note that a transaction may be aborted due to a collision, with probability v given by (6), or it may be aborted because it finds itself on the arbiter's hit list. Denoting the probability of the latter occurrence by β, we can write

$$R = \lambda[v + (1 - v)\beta]. \tag{21}$$

To find an expression for the probability β, note that a transaction, A, is aborted by the arbiter if (i) A goes to the arbiter and (ii) another successfully committing transaction, B, which arrived at the arbiter before A, had A in its list of predecessors (A would then have been added to the hit list). That is, B arrives in the system during the run time of A, tries to update one of the edges that A has updated, completes before A, goes to the arbiter and is allowed to commit.

Suppose that A contains k updates, and let t be the instant when the j-th of those updates is attempted. The average interval from t until the completion of A, given that all updates succeed, is $(k + 1 - j)b$. If the j-th update is of type i, let h_i be the average interval from t until the completion of the next transaction, B, that updates the same edge and then goes to the arbiter. That average can be estimated as

$$h_i = \frac{1}{\xi_i \alpha} + \frac{\kappa - r_1}{2(1 - r_1)} b, \tag{22}$$

where α is given by (18). The multiplier of b in the right-hand side is half of the average number of updates in a transaction, given that there are more than one.

Denote by β_{ijk} the probability that B arrives after the j-th update out of the k in A, and completes before A, and A goes to the arbiter but is aborted because B commits, given that the j-th update is of type i. We write

$$\beta_{ijk} = \alpha(1 - \beta) \frac{(k + 1 - j)b}{h_i + (k + 1 - j)b}. \tag{23}$$

Removing the conditioning on the type of update, we get the probability, β_{jk}, that A is aborted by the arbiter due to the j-th of its k updates:

$$\beta_{jk} = \sum_{i=1}^{T} \beta_{ijk} p_i. \tag{24}$$

The probability, β_k, that at least one of the k updates will cause A to be aborted, is

$$\beta_k = 1 - \prod_{j=1}^{k} (1 - \beta_{jk}). \tag{25}$$

Finally, the unconditional probability, β, that an arbitrary transaction is aborted by the arbiter, can be expressed as

$$\beta = \sum_{k=2}^{\infty} \beta_k r_k. \tag{26}$$

The right-hand side of this equation depends on α, which has already been computed, and also on β. Thus, we have another fixed-point equation which can be solved by an iterative procedure of the type (20).

One might wish to measure the performance of the system by a cost function of the form

$$C = c_1 W + c_2 R, \tag{27}$$

where c_1 and c_2 are some coefficients reflecting the relative importance given to the average response time and number of aborts. There are trade-offs that may need to be controlled. If, for example, the arbiter is overloaded, leading to large or infinite response times, a 'voluntary abort' policy may be introduced. If a transaction cannot commit upon completion (because its predecessor list is non-empty), it tosses a biased coin and, with probability σ, aborts instead of going to the arbiter. The offered load at the arbiter queue would then be reduced to $\rho = \lambda\alpha(1 - \sigma)s$. The optimal value of σ would be chosen so as to minimize the cost function C.

5   Numerical and Simulation Results

The purpose of this section is to assess the accuracy of the model estimates by comparing them with simulations. In order to reduce the number of parameters to be set, we focus on the smallest and most frequently accessed class of edges, ignoring the larger classes where conflicts are very unlikely to occur. The examples we have chosen contain a single class with N distributed edges, each of which is equally likely to be the target of an update. The size and traffic parameters are typical of a large scale-free graph database (see also [5]).

In the first example, N is varied between 5000 and 25000 edges. The arrival rate is fixed at $\lambda = 1000$ transactions per second. The average network delay is assumed to be $5\,\mathrm{ms}$ (i.e., $\delta = 200$). That is also the value of b (the average time per update). The distribution of the number of updates in a transaction is geometric, with mean $\kappa = 5$. The average arbiter service time is $s = 0.01$ and that value will be kept fixed in the following examples.

In Fig. 3, the total average number, R, of transaction aborted per unit time by the collision detection and by the order arbitration parts of the protocol, is plotted against the number of edges. The estimated points are computed by the algorithm described in Sect. 3, while each simulated point represents the result of a simulation run where one million transactions pass through the system.

Intuitively, we expect that when the number of edges increases, there will be fewer collisions and instances of overtaking, and therefore fewer aborts. Indeed, that is what is observed. The model consistently underestimates the number of aborts, but the relative errors are not large. They vary from 9% at $N = 5000$ to 5% at $N = 25000$. That underestimation is probably caused by the simplifying assumptions used in deriving the approximate estimates. On the other hand, the times taken to produce the two plots were vastly different: the model plot took a small fraction of a second to compute, while the simulation runs were several orders of magnitude slower.

From now on, the number of edges will be fixed at $N = 10000$ and the effect of different parameters will be explored. In the second example, the arrival rate λ is varied between 700 and 1200 transactions per second, while the other parameters are kept as before.

In Fig. 4, the average number of aborted transactions per second, R, is plotted against the arrival rate λ, using both the model approximation and simulations. Each simulated point is again the result of a run where one million transactions

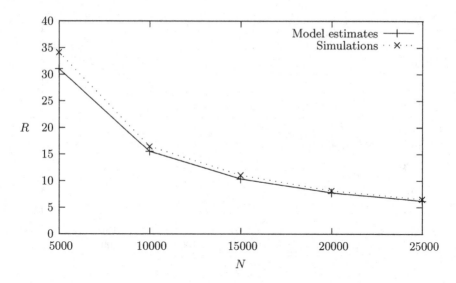

Fig. 3. Abort rate as a function of N $\lambda = 1000$, $\kappa = 5$, $\delta = 200$, $b = 0.005$, $s = 0.01$

pass through the system. Once more, we observe that the model slightly underestimates the values of R, but the relative errors are quite small; they are on the order of 6% or less, over the entire range.

The average response time of a transaction, W, was about 25 ms; its value changed very little over this range of arrival rates.

For these parameter values, the model predicts that the arbiter queue becomes unstable when the arrival rate is about $\lambda = 1500$. The simulation agrees. The observed rate at which transactions join the arbiter queue exceeds the service rate, $\mu = 100$, for that value of λ.

For the next experiment, the average network delay is doubled to 10 ms, $\delta = 100$. Intuitively, this should have the effect of increasing the rate at which transactions are aborted, and also should increase the offered load at the arbiter queue.

Figure 5 confirms our intuition. The relative errors of the model estimates are still quite low, on the order of 9% or less. The arrival rate is now varied between $\lambda = 600$ and $\lambda = 1000$. Both the model and the simulation agree that the arbiter queue becomes unstable when $\lambda = 1100$.

In the fourth experiment, the network delay is back to 5 ms, but the number of updates in a transaction, K, has a different distribution and mean. The assumption now is that K is uniformly distributed on the range [1,19], with a mean of 10. The results are illustrated in Fig. 6

The larger number of updates per transaction leads to both higher likelihood of collisions and more visits to the arbiter. The saturation point for the arbiter queue is now a little below $\lambda = 550$. As Fig. 6 illustrates, the model approximation is still accurate, with relative errors on the order of 8% or less.

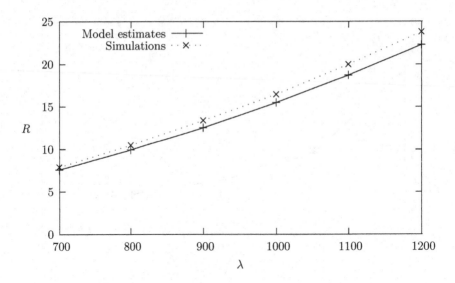

Fig. 4. Abort rate as a function of λ $\kappa = 5$, $\delta = 200$, $b = 0.005$, $s = 0.01$

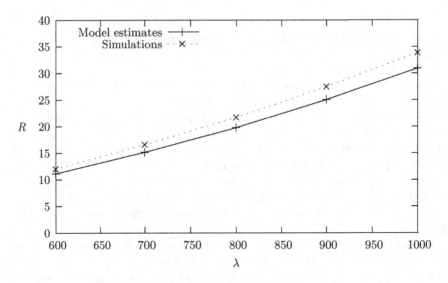

Fig. 5. Larger network delays $\kappa = 5$, $\delta = 100$, $b = 0.01$, $s = 0.01$

It is perhaps worth noting that in the last three examples, the rate of aborts increases roughly linearly with λ. For all arrival rates in example 2, between 1% and 2% of the incoming transactions are aborted. In example 3 that fraction is between 2% and 3%, while in example 4 it is between 3% and 4%.

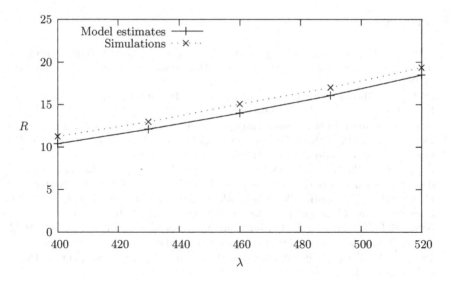

Fig. 6. Different distribution of updates $\kappa = 10$, $\delta = 200$, $b = 0.005$, $s = 0.01$

6   Conclusion

We have addressed the information corruption problem caused by interferences among transactions which update distributed edges. The proposed concurrency control protocol has two distinct mechanisms: collision detection and arbitration between transactions. That protocol has an impact on system performance, in terms of aborted transactions and load on the arbiter. To evaluate this impact, an approximate model was developed and solved. It provides estimates for the average number of transactions that are aborted per unit time, the probability that a transaction will need to go to arbitration, and the average response time of a transaction. The accuracy of the solution was examined by comparisons with simulations and was found to be very high under a variety of parameter settings.

Similar to the edge-order inconsistency examined here, there may also be *node-order* inconsistency, occurring when transactions interfere while updating the same set of nodes. Eliminating node-order inconsistencies will be addressed in future work. It is well known in the database literature that there is a hierarchy of approaches which achieve various degrees of concurrency control (see [1]). Selecting an approach for a given application typically involves a trade-off between consistency requirements and performance. The most stringent common form of concurrency control is *serializability*, which maintains an abstraction of transactions being executed in some serial order. That requirement incurs the highest performance overhead.

References

1. Adya, A.: Weak consistency: a generalized theory and optimistic implementations for distributed transactions. Ph.D. thesis, Massachusetts Institute of Technology (1999)
2. Bailis, P., Ghodsi, A.: Eventual consistency today: limitations, extensions, and beyond. Queue **11**(3), 20–32 (2013)
3. Apache Cassandra. http://cassandra.apache.org/. Accessed 11 Dec 2019
4. DeCandia, D., et al.: Dynamo: Amazon's highly available key-value store. SIGOPS Oper. Syst. Rev. **41**(6), 205–220 (2007)
5. Ezhilchelvan, P., Mitrani, I., Webber, J.: On the degradation of distributed graph databases with eventual consistency. In: Bakhshi, R., Ballarini, P., Barbot, B., Castel-Taleb, H., Remke, A. (eds.) EPEW 2018. LNCS, vol. 11178, pp. 1–13. Springer, Cham (2018). https://doi.org/10.1007/978-3-030-02227-3_1
6. Kung, H.T., Robinson, J.T.: On optimistic methods for concurrency control. ACM Trans. Database Syst. (TODS) **6**(2), 213–226 (1981)
7. Lamport, L.: The part-time parliament. ACM Trans. Comput. Syst. (TOCS) **16**(2), 133–169 (1998)
8. Ongaro, D., Ousterhout, J.: In search of an understandable consensus algorithm. In: USENIX Annual Technical Conference, pp. 305–319. USENIX Association, Philadelphia (2014)
9. Escriva, R., Wong, B., Sirer, E.G.: Warp: lightweight multi-key transactions for key-value stores. CoRR:abs/1509.07815 (2015)
10. Huang, J., Abadi, D.J.: Leopard: lightweight edge-oriented partitioning and replication for dynamic graphs. VLDB Endow. **9**(7), 40–551 (2016)
11. Firth, H., Missier, P.: TAPER: query-aware, partition-enhancement for large, heterogeneous graphs. Distrib. Parallel Databases **35**(2), 85–115 (2017)
12. Robinson, I., Webber, J., Eifrem, E.: Graph Databases, New Opportunities for Connected Data. O'Reilly Media, Inc., Sebastopol (2015)
13. Stanton, I., Kliot, G.: Streaming graph partitioning for large distributed graphs. In: 18th ACM SIGKDD International Conference on Knowledge Discovery and Data Mining, pp. 1222–1230. ACM, Beijing (2012)
14. Vogels, W.: Eventually consistent. Comm. ACM **52**(1), 40–44 (2009)

A Novel Data-Driven Algorithm for the Automated Detection of Unexpectedly High Traffic Flow in Uncongested Traffic States

Bo Klaasse[1], Rik Timmerman[1]([✉]) [iD], Tessel van Ballegooijen[2], Marko Boon[1] [iD], and Gerard Eijkelenboom[2]

[1] Eindhoven University of Technology, Eindhoven, The Netherlands
r.w.timmerman@tue.nl
[2] De Verkeersonderneming, Rotterdam, The Netherlands

Abstract. We present an algorithm to identify days that exhibit the seemingly paradoxical behaviour of high traffic flow and, simultaneously, a striking absence of traffic jams. We introduce the notion of high-performance days to refer to these days. The developed algorithm consists of three steps: step 1, based on the fundamental diagram (i.e. an empirical relation between the traffic flow and traffic density), we estimate the critical speed by using robust regression as a tool for labelling congested and uncongested data points; step 2, based on this labelling of the data, the breakdown probability can be estimated (i.e. the probability that the average speed drops below the critical speed); step 3, we identify unperturbed moments (i.e. moments when a breakdown is expected, but does not occur) and subsequently identify the high-performance days based on the number of unperturbed moments. Identifying high-performance days could be a building block in the quest for traffic jam reduction; using more detailed data one might be able to identify specific characteristics of high-performance days. The algorithm is applied to a case study featuring the highly congested A15 motorway in the Netherlands.

Keywords: High-performance days · Traffic breakdown · Data-driven algorithm · Fundamental diagram · Congestion · Detector data

1 Introduction

Nowadays, traffic jams have become an inevitable part of road traffic. In particular, near or in urban areas the high vehicle-to-capacity ratio on the road imposes cars to slow down or even stop too frequently. This causes a wide variety of problems, as in the Netherlands alone, the amount of monetary value lost due to traffic jams in 2018 is estimated at 1.3 billion euros [3]. Moreover, traffic jams cause pollution and decrease the quality of life e.g. in cities.

© Springer Nature Switzerland AG 2020
M. Gribaudo et al. (Eds.): EPEW 2019, LNCS 12039, pp. 65–83, 2020.
https://doi.org/10.1007/978-3-030-44411-2_5

Reducing traffic congestion is a challenging problem. Obviously, increasing the capacity of existing roads, e.g. by adding lanes, would provide a solution to the problem. However, such actions are costly and not always desired, or even possible. Alternatively, one could aim to influence drivers' behaviour. This can be achieved by, for example, monetary means (such as toll systems or congestion pricing, see e.g. [2,7]), encouraging drivers to drive outside peak hours (see [5] for instance) or dynamic road signalling (see e.g. [8]). It is increasingly important to find the exact effect of these measures, but this is a complicated problem, which is partly due to the highly complex nature of traffic and the fact that the manifestation of congestion is subject to randomness, see for example [1,20].

In this paper, we approach the problem of reducing traffic congestion from a different perspective, as we look at the *absence* of traffic jams. Typically, once the traffic flow, i.e. the throughput measured in vehicles per hour, has passed a certain threshold, congestion *could* emerge. This phenomenon is referred to as a "breakdown". We are interested in days during which a relatively large number of breakdowns were expected, but did not occur. Such days will be referred to as "high-performance days". Specifically, we develop an algorithm to automatically identify these high-performance days based on historical traffic data and test our method on a section of the A15 motorway in the Netherlands. In a future study, one could try to determine the specific characteristics of the resulting high-performance days using more detailed data. Ultimately, the goal is to find out whether the high-performance days could be caused by specific behavioural patterns of individual drivers. However, we focus on the first step, namely the automated detection of high-performance days.

Our algorithm relies on the shape of the fundamental diagram, the well-known empirical diagram that displays the relationship between the traffic flow q (vehicles per hour) and the traffic density ρ (vehicles per kilometre) at a specific location. Many studies have shown that the fundamental diagram can be divided into two regions, a region for congestion and a region for free flow. The empirical fundamental diagram has been studied extensively and a wide variety of theoretical models has been proposed (see for example [6] for an overview). Our aim is not a theoretical model for the fundamental diagram however; we are merely interested in the critical speed, i.e. the speed which defines congestion and separates the free-flow region from the congestion region in the fundamental diagram. So, we can get around the problem of modelling the congestion region and exploit the roughly linear flow-density relationship during free flow. We show that *robust regression* is an excellent technique to obtain the free-flow speed and subsequently distinguish between free flow and congestion based on the calculated weights. Utilizing the method proposed by [1], we subsequently estimate the *breakdown probability*. This paves the way to identifying high-performance days.

To the best of our knowledge, our approach to obtain the critical speed and the introduction of the notion of high-performance days are original. Many papers focus on (real-time) traffic jam estimation using GPS-data and/or trajectory data, see a.o. [15,16,21]. This is partly due to the widespread availability of GPS data. However, we have chosen to use detector data, as traffic detectors are present on most Dutch motorways and provide a sufficiently high granularity. Detector data is also used in the literature; in [12] detector data is used to automatically track congestion and in [10] detector data is used to study phase transitions on German highways. However, the work that is probably closest to our study is [4]. Therein, the authors use detector data to estimate highway characteristics such as the free-flow speed and the critical density. These quantities are then used to calibrate a cell transmission model. We determine a related highway characteristic (the critical speed), but in our study this is a tool to estimate the breakdown probability. Indeed, our main goal is different: we identify a surprising *absence* of traffic jams. This could be an important first step towards a better understanding of the reasons why on certain days the traffic flow is so much better than on other days, although the circumstances seem to be identical.

In Sect. 2 we provide information about the location of the experimental region and discuss the data. We proceed with the theoretical foundation and the three main steps of the algorithm in Sect. 3. The validation of important assumptions and parameter choices is presented in Sect. 4, as well as the main insights of the case study. We close with a conclusion and multiple suggestions for future research in Sect. 5.

2   Description of the Location and the Data

In this section, we discuss the relevant aspects of the part of the A15 motorway from which the data is obtained. Subsequently, we elaborate on the structure of the data set and which steps we take in the preprocessing of the data.

2.1   Location of the Experimental Region

The location under consideration is the A15 motorway near Rotterdam, at the interchange with Papendrecht (see Fig. 1). Five detectors have been placed in the eastern direction, with a distance of approximately 300 m between consecutive detectors (see Fig. 1b). Between the second and third detector, an off-ramp to Papendrecht is located. Shortly afterwards, the vehicles on the A15 merge from three to two lanes. The maximum speed along this whole trajectory of the A15 is 120 km/h. The traffic jams on this trajectory belong to the most costly traffic jams in the Netherlands (see [3]) and the A15 is one of the most congested roads in the Netherlands, connecting one of the world's largest ports with the European main land, which makes this a particularly interesting motorway to study.

2.2   Description of the Data Set

The data is obtained from the Dutch National Data Warehouse for Traffic Information (NDW), a collaboration of 19 public authorities that cooperate on collecting, storing, and redistributing data. The data is publicly available and can be

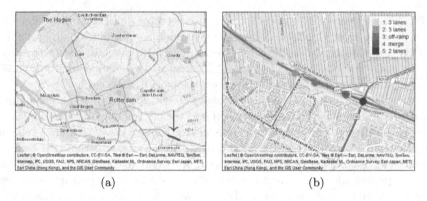

Fig. 1. (a) Overview of the trajectory, marked red and indicated by the red arrow, in relation to Rotterdam. (b) The location of the five detectors on the trajectory. (Color figure online.)

requested at the website of the NDW [14]. The data we obtained from the NDW spans a period from January 1, 2018 until December 31, 2018. Every minute, the detectors measure, for each lane individually, the number of vehicles that have passed (i.e the traffic flow q, in vehicles per hour) and the average speed v of the passing cars in kilometres per hour, calculated using the arithmetic mean. We can estimate the average traffic density ρ using $\rho = q/v$, although this formula is known to underestimate the density when the arithmetic mean is used [11]. We combine the various lanes as in [19]. For the sake of reducing the variability in the data, we aggregate the measurements to a period of 5 min, as is done in [1]. The arithmetic mean is used to obtain the average traffic flow and the average speed is calculated analogous to the average speed over multiple lanes.

The resulting data set can be described as follows. We introduce the set of locations $\mathcal{I} := \{1, 2, 3, 4, 5\}$, in accordance with Fig. 1b. Moreover, we focused our research on weekdays and thereby excluded all weekend days from the data, because the traffic flow is oftentimes significantly lower. The set containing all 261 weekdays in 2018 is denoted by $\mathcal{J}$. After the aforementioned exclusions, we have one set of measurement dates $\mathcal{J}^{(i)} \subseteq \mathcal{J}$ for each detector $i \in \mathcal{I}$. At each location we have measurements of the average traffic flow and average vehicle speed, as well as an estimate for the density, aggregated to 5-minute intervals. Hence, for location $i \in \mathcal{I}$ and date $j \in \mathcal{J}^{(i)}$ we have a sequence of measurement times

$$\mathcal{T}^{(i,j)} := \left\{ t_1^{(i,j)}, t_2^{(i,j)}, \ldots \right\} \subseteq \mathcal{T}, \tag{1}$$

where $\mathcal{T}$ is the set containing all 5-minute intervals on a day. The corresponding set of measurements for detector i on date j is

$$\mathcal{X}^{(i,j)} := \left\{ \left(q_t^{(i,j)}, v_t^{(i,j)}, \rho_t^{(i,j)} \right) : t \in \mathcal{T}^{(i,j)} \right\}. \tag{2}$$

The data set containing only the flow and the density is denoted by

$$\bar{\mathcal{X}}^{(i,j)} := \left\{ \left(\rho_t^{(i,j)}, q_t^{(i,j)} \right) : t \in \mathcal{T}^{(i,j)} \right\}. \tag{3}$$

In total we have $|\mathcal{I}| = 5$ locations and $|\mathcal{J}| = 261$ dates, leading to a total of $5 \cdot 261 = 1305$ instances. However, in the first step of the algorithm (i.e. estimating the critical speed), we do not include all days/critical speeds:

1. We exclude the most extreme critical speeds of each location (see Sect. 3 for a motivation in relation to our assumptions and Eq. (11)/the last paragraph of Sect. 3.2 for a further elaboration);
2. We exclude instances where the free-flow and congestion region are not linearly separable by a straight line through the origin, given the labelling (see Remark 2);
3. We exclude days with little or no congestion (see Sect. 4.2).

For the remaining steps, we do include all 1305 instances, meaning that no weekdays are beforehand excluded when identifying the high-performance days.

All the analyses were performed in the statistical software package R.

3   The Main Algorithm

We present the main algorithm in this section and elaborate on the theoretical foundation using traffic theory, robust regression and the estimator for the breakdown probability proposed in [1]. The algorithm consists of three parts: (i) estimating the critical speed, (ii) estimating the breakdown probability, and (iii) identifying the high-performance days. In Sect. 3.1 we formally define the relevant notions, such as the critical speed. In Sect. 3.2 we explain how the critical speed is obtained using robust regression as a labelling tool. Lastly, in Sect. 3.3 we discuss the estimator for the breakdown probability and provide a definition for high-performance days, based on "unperturbed moments".

3.1   The Fundamental Diagram and the Critical Speed

Studying the traffic behaviour at a specific location, say location i, one can distinguish two different traffic states: free flow and congestion. As in [9], we can define free flow and congestion based on the critical speed.

Definition 1 (Free flow, Congestion and Critical speed). Free (traffic) flow *is a state when the vehicle density in traffic is small enough for interactions between vehicles to become negligible. Therefore, vehicles have an opportunity to move at their desired maximum speeds [9]. When the density increases beyond a certain threshold in free flow, vehicle interaction cannot be neglected anymore. Due to this vehicle interaction, the average vehicle speed decreases to a value lower than the* critical speed, *which is the minimum average speed that is still possible in free flow. This new state of traffic is referred to as a state of* congested traffic.

We denote the critical speed at location i by $v_{\text{crit}}^{(i)}$. In the fundamental diagram, this critical speed separates the free-flow region from the congestion region. The free-flow set of location i on date j, i.e. the set containing all data points corresponding to free flow, is defined as

$$\mathcal{F}^{(i,j)} := \left\{ \left(q_t^{(i,j)}, v_t^{(i,j)}, \rho_t^{(i,j)} \right) \in \mathcal{X}^{(i,j)} : v_t^{(i,j)} \geq v_{\text{crit}}^{(i)} \right\}, \qquad (4)$$

i.e. the set of all data points of location i and date j for which the average speed is equal to or higher than the critical speed of location i. Naturally, the congestion set is defined as the complement of the free-flow set, i.e.

$$\mathcal{C}^{(i,j)} := \mathcal{X}^{(i,j)} \setminus \mathcal{F}^{(i,j)}. \qquad (5)$$

The difference between free flow and congestion is clearly visible in the fundamental diagram (or the empirical fundamental diagram of traffic flow, to be precise), which is a plot of the measured flow rates $q_t^{(i,j)}$ against the vehicle densities $\rho_t^{(i,j)}$. An example of the empirical fundamental diagram is presented in Fig. 2a. In this example, the black line clearly separates the free flow set from the congestion set.

During free flow, the flow-density relationship can be modelled by a straight line (see the orange line in Fig. 2a), which logically must pass through the origin:

$$q \approx \rho \cdot v_{\text{free}}^{(i)} \quad \forall (q, v, \rho) \in \mathcal{F}^{(i,j)}. \qquad (6)$$

When using the data set $\mathcal{X}^{(i,j)}$, we assume the following conditions are met:

(i) The average speed during free flow $v_{\text{free}}^{(i)}$ is constant for all locations $i \in \mathcal{I}$;
(ii) The road conditions at location i are homogeneous for all dates $j \in \mathcal{J}^{(i)}$, for all locations $i \in \mathcal{I}$;
(iii) For each $i \in \mathcal{I}$ and $j \in \mathcal{J}^{(i)}$, the number of free-flow measurements significantly exceeds the number of congestion measurements.

Whenever at least one of these conditions is violated, for a certain day j at location i, day j will not be taken into account when determining $v_{\text{crit}}^{(i)}$. The first condition is rarely violated, since a constant free flow speed follows from the definition of free flow (see e.g. [9]), given conflict free roads with a fixed speed limit and homogeneous conditions. Assumptions (ii) and (iii) may be violated on days where circumstances are completely different from ordinary days, for example in case of accidents, road works or extreme weather conditions. These days could be detected using additional data and therefore be removed from the data set. However, in order to keep the algorithm as simple and self-contained as possible, we simply choose to exclude the most extreme critical speeds. We emphasise that in our experimental region the core elements of the road were fixed throughout the year, i.e. the speed limit is fixed and no traffic lanes where removed or added. Furthermore, despite the experimental region being subject to heavy congestion, congestion occurs mainly during the morning and afternoon rush hour, which means that in general the number of free-flow measurement well exceeds the number of congestion measurements. As a result, Assumptions (i), (ii) and (iii) are only violated in extreme cases and removing the most extreme critical speeds will be sufficient to ensure the assumptions are met. This explains the first point regarding the removal of several critical speeds stated in Sect. 2.2.

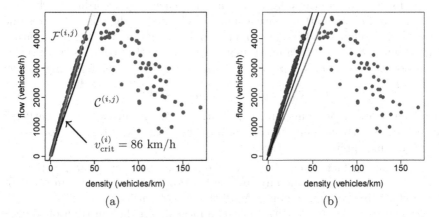

Fig. 2. The fundamental diagram with free-flow points (green) and congestion points (red). In (a) it is shown how the free-flow region and the congestion region are linearly separable by a straight line through the origin (the black line), the slope of this line is the critical speed. Additionally, the slope of the orange line through the origin is the (constant) free-flow speed, which is 95.5 km/h. Note that the free-flow speed is significantly below the speed limit, as this is an average over both multiple vehicles and multiple lanes. In (b) it is shown how the critical speed can be estimated by the line that lies exactly between the boundary line of the free-flow region (blue) and the boundary line of the congestion region (magenta). (Color figure online.)

3.2   Using Robust Regression to Label Data Points

The purpose of our algorithm is to find the free flow set and the congestion set, for every day and location. More formally, we aim to find a label for each $(q, v, \rho) \in \mathcal{X}^{(i,j)}$ that indicates whether $(q, v, \rho) \in \mathcal{F}^{(i,j)}$ or $(q, v, \rho) \in \mathcal{C}^{(i,j)}$. A logical first step is to determine the straight line through the origin that lies exactly between the free-flow region and the congestion region, as depicted by the black line in Fig. 2b. The slope of this line is the estimate of the critical speed of location i for each date $j \in \mathcal{J}^{(i)}$, denoted by $v_{\text{crit}}^{(i,j)}$.

In order to obtain the critical speed and the corresponding labelling from the fundamental diagram, several methods have been studied in the literature. Examples are an iterative regression method after performing a change-point analysis [1], the use of fuzzy logic for clustering [17], and assuming a specific model for the fundamental diagram, obtaining the critical density and subsequently labelling each point [11]. However, we opt for a more intuitive and efficient method based on robust regression, to exploit the underlying structure of the fundamental diagram.

Robust regression essentially does the same as ordinary regression, yet is more robust to potential violations of the modelling assumptions (e.g. outliers), see for example [13]. To this end, each data point $\mathbf{x}$ is assigned a weight $w(\mathbf{x}) \in [0, 1]$ and subsequently a linear model is fitted and a reiterative weighted least squares fit is performed (where the weights are updated each step according to the new estimate); in this way outliers have a smaller influence on the final estimates due to their lower weights and the model aims to fit the majority of the data, rather than the whole data set. We apply robust regression to the flow-density set $\bar{\mathcal{X}}^{(i,j)}$ of each location i and date j separately. Specifically, we fit the following model:

$$q_t = v_{\text{free}}^{(i,j)} \cdot \rho_t + \varepsilon_t \quad \forall (\rho_t, q_t) \in \bar{\mathcal{X}}^{(i,j)}, \tag{7}$$

where the ε_t are error terms with expectation zero. In our case, the "outliers" are the points corresponding to congestion. The reason why this method works so well for this application, is threefold:

1. We exploit the fact that in free flow, the relation between q and ρ is linear;
2. We do not have to assume any specific relation between q and ρ in the congested set, because these points fulfill the role of outliers;
3. The method computes weights that are a measure for the contribution of each point to the final estimate, which can be used for the labelling.

Remark 1. Assumption (iii) from Subsect. 3.1, specifying that we only consider days where the number of points corresponding to congestion is smaller than the number of free flow points, is essential. On a day where this assumption is violated, we have more points belonging to congestion, meaning that the fitted regression line would no longer pass through the free flow set. In this case, the estimated free flow speed $v_{\text{free}}^{(i,j)}$ would be significantly lower than the maximum speed, which makes these days extremely easy to detect (and remove).

The robust regression is performed using the function rlm from the MASS-package in R, with MM-estimation and Tukey's Bisquare function for the weights with the default S-estimator as suggested in [22]. Tukey's Bisquare function behaves similarly to the squared error function except for larger errors, for which it decreases the weight (see e.g. [13]). This results in an estimate for $v_{\text{free}}^{(i,j)}$ and certain weights $w(\mathbf{x})$ for each data point $\mathbf{x} \in \bar{\mathcal{X}}^{(i,j)}$. Instead of the usual interest in the model and parameter estimation, we are interested in the *weights* associated with each data point. Using the weights, we perform the labelling: if the weight is low and if the data point corresponds to a speed lower than the free-flow speed, $v_{\text{free}}^{(i,j)}$, the data point will be labelled as congestion. All other points will be labelled as free flow. Hence, for each $\mathbf{x} = (\rho, q) \in \bar{\mathcal{X}}^{(i,j)}$ we determine $\mathbb{1}_C(\mathbf{x}) := \mathbb{1}\{\mathbf{x} \equiv (q, v, \rho) \in C^{(i,j)}; \mathbf{x} \in \bar{\mathcal{X}}^{(i,j)}\}$, i.e. the indicator function for the event that $\mathbf{x}$ corresponds to congestion or not. The critical weight has been placed at 0.01 (see Sect. 4.2 for a justification), hence

$$\mathbb{1}_C(\mathbf{x}) = \begin{cases} 1 & \text{if } w(\mathbf{x}) < 0.01 \text{ and } v = q/\rho < v_{\text{free}}^{(i,j)} \\ 0 & \text{otherwise.} \end{cases} \tag{8}$$

After we obtain the labels, we estimate $v_{\text{crit}}^{(i,j)}$ (see the black line in Fig. 2b) by determining the slope of the straight line through the origin that lies exactly between the free-flow region and the congestion region.

Remark 2. It may happen that the boundary line of the congestion region lies above the boundary line of the free-flow region (i.e. the magenta line has a larger slope than the blue line in Fig. 2b), since the weights are calculated based on the Euclidean distance from the free-flow line. In this case, the free-flow region and the congestion region are not linearly separable by a straight line through the origin, given the labelling. For such instances, there will exist data points $\mathbf{x} \equiv (q, v, \rho) \in \mathcal{C}^{(i,j)}$ and $\mathbf{x}' \equiv (q', v', \rho') \in \mathcal{F}^{(i,j)}$ such that $v > v'$. The critical speed for such instances is indeterminate and therefore we do not include these instances in the determination of the critical speed of the corresponding location.

In the end, the critical speed of location i is estimated as follows:

$$v_{\text{crit}}^{(i)} = \text{median}\{\mathcal{V}_{\text{crit}}^{(i)}\}, \tag{9}$$

where

$$\mathcal{V}_{\text{crit}}^{(i)} := \left\{ v_{\text{crit}}^{(i,j)} \right\} \tag{10}$$

such that:

$$\left| v_{\text{crit}}^{(i,j)} - \mu\{v_{\text{crit}}^{(i,j)}\}_{j \in \mathcal{J}^{(i)}} \right| < 2\sigma\{v_{\text{crit}}^{(i,j)}\}_{j \in \mathcal{J}^{(i)}}; \tag{11}$$

$$v' > v \quad \forall \mathbf{x} \in \mathcal{C}^{(i,j)}, \mathbf{x}' \in \mathcal{F}^{(i,j)}; \tag{12}$$

$$\text{MAPE}\left(\bar{\mathcal{X}}^{(i,j)}\right) \geq 0.1. \tag{13}$$

where $\mu\{\cdot\}$ and $\sigma\{\cdot\}$ denote the mean and standard deviation of the corresponding sets respectively and $\text{MAPE}\left(\bar{\mathcal{X}}^{(i,j)}\right)$ denotes the mean absolute percentage error of the regression model presented in Eq. (7).

Equation (11) removes the most extreme critical speeds. By excluding days with a critical speed that lies outside a range of twice the standard deviation from the average, we prevent potential violations of the assumptions from influencing the estimates (as elaborated upon in Sect. 3.1). Equation (12) excludes days where the boundary line of the congestion region lies above the boundary line of the free-flow region (see Remark 2). Lastly, Eq. (13) ensures that the critical speed of a location is not based on days with little or no congestion. As one can imagine, in case of hardly any congestion, a free-flow point with a relatively slow speed might be incorrectly labelled as congestion. We therefore impose a minimal level of congestion and use the mean absolute percentage error (MAPE, see e.g. [18]) of the corresponding model (see Eq. (7)) as a surrogate of the average congestion level. The MAPE expresses the error of the model in terms of a percentage; a low MAPE corresponds to a very accurate model, implying hardly any congestion, whereas a high MAPE indicates that various points deviate from the straight line through the origin, which corresponds to the presence of congestion during that day. The critical level of the MAPE has been placed at 0.1, in Sect. 4.2 this threshold will be motivated.

The set of critical speeds of location i, corresponding to the instances of location i which satisfy the three conditions presented in Eqs. (11), (12) and (13), is given by $\mathcal{V}_{crit}^{(i)}$. The critical speed of location i is subsequently determined by taking the median of this set. We take the *median* of the critical speeds among multiple days to provide a solid baseline for comparison among different days. We emphasise that in the end the critical speed of each location is estimated as the median of at least 147 critical speeds (out of 261 weekdays) and that most instances were removed based on Eq. (12).

3.3 Estimating the Breakdown Probability and Identifying the High-Performance Days

Congestion arises as a consequence of a breakdown, which is defined as a transition from free flow to congestion (see, e.g. [1]). Usually, this happens when the traffic flow is high and some kind of disruption occurs (e.g. a vehicle changing lanes or another sudden movement of a driver).

Definition 2 (Breakdown). *A breakdown, at location i and date j, is a moment $t_k^{(i,j)} \in \mathcal{T}^{(i,j)}$ such that*

$$v_{t_k^{(i,j)}}^{(i,j)} \geq v_{crit}^{(i)} > v_{t_{k+1}^{(i,j)}}^{(i,j)}.$$

We assume that breakdowns have a probabilistic nature, see e.g. [1,20], meaning that from a macroscopic point of view the occurrence of breakdowns (given a certain traffic flow) is random. This implies the existence of a breakdown probability (as a function of the traffic flow). To estimate this probability, we use the non-parametric estimator discussed in Arnesen and Hjelkrem [1]. To calibrate this estimator, the aforementioned classification of each data point as either free flow or congestion is required. Arnesen and Hjelkrem define two functions; $Q^{(i)}(q)$, which is the number of breakdowns at location i while the traffic flow is equal to or lower than q, and $R^{(i)}(q)$, which is the number of times a breakdown did not occur at location i with a traffic flow of at least q. Subsequently, the breakdown probability $P^{(i)}(q)$, which denotes the probability of a breakdown at location i when the traffic flow is q, can be estimated by

$$P^{(i)}(q) = \frac{Q^{(i)}(q)}{Q^{(i)}(q) + R^{(i)}(q)}. \tag{14}$$

Remark 3. To avoid including "fake breakdowns" (e.g. a single vehicle driving unnecessarily slow at night), we pose the additional constraint on a breakdown that it does not happen before 5:00 in the morning. Indeed, multiple times we observed before 5:00, at a minimal traffic flow, a sudden drop of the average speed to just below the critical speed. We assume that such events are not relevant for estimating the breakdown distribution as this could be a truck driving at its speed limit of 80 km/h.

To reduce the complexity of the estimation method, we use a surrogate for the breakdown probabilities, obtained by fitting a cumulative normal distribution function, as is done in [1].

In Sect. 1, an intuitive description of a high-performance day was given. In this section we present a criterion to determine a quantitative definition for high-performance days. To this end, we employ the estimated breakdown probability in Eq. (14), to find *unperturbed moments*. An unperturbed moment is a moment at which the probability of a breakdown is at least 0.5, but the expected breakdown did not occur, or more mathematically:

Definition 3 (Unperturbed moment). *An unperturbed moment, at location i on date j, is a moment $t_k^{(i,j)} \in T^{(i,j)}$ with intensity $q_{t_k^{(i,j)}}^{(i,j)} \geq q_{upt}^{(i)}$ and speed $v_{t_k^{(i,j)}}^{(i,j)} \geq v_{crit}^{(i)}$ for which it holds that*

$$P^{(i)}\left(q_{t_k^{(i,j)}}^{(i,j)}\right) \geq 1/2 \quad \wedge \quad v_{t_{k+1}^{(i,j)}}^{(i,j)} \geq v_{crit}^{(i)}, \tag{15}$$

where $q_{upt}^{(i)}$ is the smallest value of the traffic flow q such that $P^{(i)}(q) \geq 1/2$.

A plausible definition of a high-performance day follows naturally.

Definition 4 (High-performance day). *A high-performance day is a day with a large number of consecutive unperturbed moments in both time and space compared to other days.*

Note that a high-performance day is thereby a relative measure, as it will depend on the location how many unperturbed moments are generally present (some locations experience more variability in terms of breakdowns in relation to the traffic flow). Indeed, a certain level of freedom in the definition of high-performance days is required. For example, quantifications such as the top 0.05 percentile, though plausible in some cases, incorrectly imply the existence of high-performance days at any location. Furthermore, concretizations of the definition in terms of the number of unperturbed moments depend on the experimental region.

4 Key Insights and Validation

In this section, we present the results of our algorithm and validate the estimation methods. In particular, we study the results of the three steps of the algorithm and present several measures of the top 10 high-performance days. In addition, we take a closer look at what exactly a high-performance day looks like and how we can use our macroscopic data to visualise the dynamics of such days for the whole trajectory. Subsequently, we elaborate on several problems one might encounter when applying the method at a different location and how these problems could be tackled. Specifically, we state how we dealt with these problems and how we obtained the critical weight and the critical level of the MAPE.

4.1   Results and Key Insights

In Table 1 we present the results of the first and second step of the algorithm (i.e. estimating the critical speed and the breakdown probabilities respectively). We observe that the critical speed is roughly equal for the various locations. We see a similar behavior for the estimated free-flow speeds, which are consistently roughly 10 km/h above the corresponding estimated critical speeds. Furthermore, we observe that the smallest value of the traffic flow for which the breakdown probability is at least 0.5 decreases along the trajectory, meaning that the last two locations experience breakdowns at a lower traffic flow than the first three locations. This makes sense considering the merge from 3 lanes to 2 lanes at the fourth location.

Table 1. Columns from left-to-right: the rounded estimated critical speed of location i, the rounded estimated free-flow speed of location i (based on the median of the free-flow speeds of the instances that were used to estimate the critical speed of location i), the number of instances used for estimating the critical speed of location i (out of a total of 261 weekdays) and the smallest traffic flow for which the breakdown probability is at least 0.5. The speeds are expressed in kilometres per hour and the traffic flows are expressed in vehicles per hour.

	$v_{\text{crit}}^{(i)}$	$v_{\text{free}}^{(i)}$	$|\mathcal{V}_{\text{crit}}^{(i)}|$	$q_{\text{upt}}^{(i)}$
Location 1	95.5	104.5	147	4358
Location 2	93	103	162	4019
Location 3	93	102	175	3901
Location 4	94.5	104.5	180	3195
Location 5	92.5	102.5	175	3164

In Fig. 3 we present a scatter plot displaying the average number of unperturbed moments per location for each weekday of 2018. Additionally, the colour of each point corresponds to the average breakdown probability of the unperturbed moments. We observe that the days can be grouped into roughly three categories: days with hardly any unperturbed moments, days with some unperturbed moments, and days with a relatively large number of unperturbed moments. It turns out that most days in the first group correspond to days with significantly less traffic, thus implying a low traffic flow and thereby a lack of unperturbed moments. For example, the grey points in Fig. 3 often correspond to (school) holidays. The third group, however, is of major interest to us, as these are the high-performance days.

In Table 2 we present several measures of the top 10 high-performance days (based on Fig. 3), corresponding to the fourth location. We choose to only present results for the fourth location, because averaging the speeds over the various locations requires a critical speed for the whole trajectory as a baseline (whose definition is not straightforward). To study the characteristics of these days, we

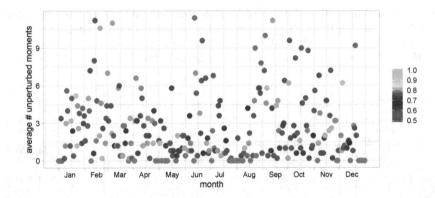

Fig. 3. Plot of the average number of unperturbed moments for each weekday of 2018. The colour of each point indicates the average breakdown probability of the unperturbed moments. In case no unperturbed moments occurred, the corresponding point is grey.

investigate the average speed and average fraction of free-flow measurements. We look at three time intervals: the morning rush hour 6.30–9.30, outside peak hours 9.30–15.30 and the afternoon rush hour 15.30–19.00. We observe that, though all days show a relatively large number of unperturbed moments, the characteristics of the various days can differ greatly. For example, the top 7 high-performance days all have an average speed during the morning rush hour that is below the critical speed of the corresponding location (i.e. 94.5 km/h) and at least 10% of the measurements during the morning rush hour correspond to congestion, whereas the remaining three days shows hardly any signs of congestion in the morning. We also observe a similar pattern across all high-performance days; the mornings are significantly better (in terms of the average speed and the fraction free flow) than the afternoons. In fact, it seems that severe congestion during the afternoon was present in almost all high-performance days (only February 14, 2018 is an exception). Nevertheless, the mornings of the top 10 high-performance days are quite extraordinary, in particular when comparing the average speed and the fraction free flow with the median over all weekdays from 2018 at location 4.

We now thoroughly study the traffic behaviour during October 17, 2018. During this day, an average of 9 unperturbed moments was identified (see Table 2). This day is particularly interesting because of the seemingly large difference between the morning rush hour and the afternoon rush hour. In fact, this day is the only day in the top 10 high-performance days which does not show any congestion during the entire morning rush hour. In Fig. 4 a joint time series of the average flow and average speed at the fourth location during this day is presented. As expected, we notice a large number of unperturbed moments, mostly during the morning. The contrast between the morning and the afternoon is indeed interesting, as the breakdown, which remained absent in the morning, manifested in the late afternoon at a lower traffic flow. This is in line with our probabilistic view on the occurrence of a breakdown, at least from a macroscopic point of view, and confirms that this morning was indeed extraordinary.

Table 2. Several measures of the top 10 high-performance days, based on Fig. 3, corresponding to the fourth location. The average speed is presented during the morning rush hour 6.30–9.30, outside peak hours 9.30–15.30 and during the afternoon rush hour 15.30–19.00, as well as the corresponding fraction free flow. The median over all weekdays of 2018 is presented as well.

	Average number unperturbed moments (per location)	Average speed morning rush hour	Average speed outside peak hours	Average speed afternoon rush hour	Fraction free flow morning rush hour	Fraction free flow outside peak hours	Fraction free flow afternoon rush hour		Average speed legend	Fraction free flow legend
12-Jun	11.4	88.5	99.3	33.2	0.76	0.96	0.12		0.0	0.00
14-Feb	11.2	92.9	99.8	75.3	0.86	0.94	0.60		10.0	0.10
13-Sep	11.2	93.4	99.2	32.1	0.83	0.94	0.07		20.0	0.20
7-Mar	11	82.9	104.2	41.9	0.72	1.00	0.36		30.0	0.30
20-Feb	10.6	58.5	97.7	52.9	0.39	0.88	0.36		40.0	0.40
4-Sep	10	95.2	104.4	41.2	0.86	1.00	0.21		50.0	0.50
21-Jun	9.6	90.3	99.2	18.3	0.81	0.96	0.00		60.0	0.60
3-Oct	9.6	99.7	103.5	30.9	0.94	1.00	0.05		70.0	0.70
20-Dec	9.2	100.8	92.2	30.8	0.97	0.86	0.12		80.0	0.80
17-Oct	9	103.1	99.5	39.3	1.00	0.96	0.19		90.0	0.90
									>94.5	1.00
Median	2.2	71.2	99.4	44.0	0.51	0.94	0.21			

Additionally, one could employ visualizations to investigate the whole trajectory simultaneously, see Fig. 5. We verified that the morning of October 17, 2018 was extraordinary at the fourth location and Fig. 5 shows that this was the case for the whole trajectory. Indeed, we observe multiple unperturbed moments during the morning rush hour at each of the five locations. In particular, despite the high traffic flow (recall that unperturbed moments only occur at a traffic flow of at least 3164 vehicles per hour, see Table 1), we observe no significant speed decrease. Furthermore, as we expect based on Fig. 4, a breakdown along the whole trajectory can clearly be seen around 15.20-15.30 (see Fig. 5).

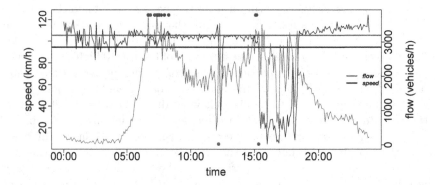

Fig. 4. Time series of the average speed (black) and average flow (red) during October 17, 2018 at location 4. Unperturbed moments are indicated by a green dot and breakdowns are indicated by a red dot. The horizontal black line is the estimated critical speed and the horizontal red line is the smallest traffic flow for which the breakdown probability is at least 0.5. (Color figure online.)

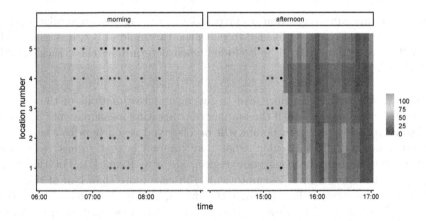

Fig. 5. A space-time diagram of the morning rush hour and the afternoon of October 17, 2018. The average speed is displayed along the whole trajectory. Furthermore, breakdowns are marked with a black marker and unperturbed moments are marked with a red dot. (Color figure online.)

4.2 Validation

The critical speeds are estimated based on a labelling of the data points resulting from the robust regression method discussed in Sect. 3.2. As the exact shape of the fundamental diagram depends on the location, it is difficult to make general statements about the accuracy of the critical speed estimation. However, we can identify three possible issues: 1. little or no congestion occurred during a day; 2. extreme congestion occurred during a day; 3. the free-flow speed was not (approximately) constant. We also present a way to determine whether or not those problems did arise (besides additional information about the experimental region). Finally, we conclude this section with a discussion on how to choose the critical weight, which is used to determine whether observations belong to the congestion set or the free-flow set.

Little or No Congestion. In this case, robust regression might interpret a free-flow point with a relatively slow speed as an outlier and therefore cause a free-flow point to be labelled as a congestion point. This leads to a higher estimate of the critical speed during that day. Though in our case it is not likely that the final estimate of the critical speed will be strongly influenced by several overestimates (considering that our experimental region is generally subject to heavy congestion), we still exclude days with little or no congestion. As mentioned in Sect. 3.2, we use the mean absolute percentage error (MAPE) of the robust regression model presented in Eq. (7) as a surrogate for the average congestion level. In Fig. 6 a scatter plot of the MAPE for the various days of location 1 is shown. We observe that, for example, during the holidays (the beginning of January/end of December) and throughout the summer break, the MAPE is close to zero. Indeed, during those days the traffic flow was significantly lower and therefore hardly any congestion occurred. Based on Fig. 6 (and similar figures for the other locations), we decided to place the threshold at 0.1; instances with a MAPE of less than 0.1 will be excluded when determining the critical speed, as in Eq. (13).

Extreme Congestion. This may lead to severe underestimations of the critical speed. One can imagine that if the number of congestion measurements becomes too large, not all congestion points will be observed as outliers by the robust regression method. In particular, what may happen is that robust regression fits a model through the congestion region, see also Remark 1. For the MM-estimators it is known that (asymptotically) in case more than half of the data points lie on a straight line through the origin, the final model will fit that line [23]. This means that, if we assume a constant flow-density relation in free flow, the free-flow speed should be accurately estimated if more than half of the measurements correspond to free-flow. However, because Eq. (6) is only an approximate relation, the algorithm will be even more sensitive to a larger congestion set. In our case study, the fraction of free flow was generally well above 0.5. However, before employing robust regression to determine the critical speed, it is recommended one verifies that the average free-flow level is above 0.5. In case the congestion level is around 0.5 one should cautiously verify that the critical speed is correctly estimated (by e.g. studying the distribution of the estimated critical speed for the various days).

Non-constant Free-Flow Speed. In case the free-flow speed is not constant, the structure of the fundamental diagram will change heavily (in comparison with e.g. Fig. 2). One example would be a decrease of the speed limit when the rush-hour lane is open. In case the rush-hour lane is opened during peak hours, this could result in a free-flow curve, rather than a straight line, displaying an average speed decrease at high traffic flows. Such a scenario could be problematic for our algorithm, as the approximate flow-density relationship, presented in Eq. (6), no longer holds. We suggest that one beforehand verifies that the free-flow speed is constant, either by using information about the experimental region or by studying the fundamental diagram. In our case there was no dynamic speed limit and the fundamental diagrams showed no indication of a non-constant free-flow speed.

Critical Weights. In Sect. 3.2, we introduced the critical weight, which is used to distinguish between congestion and free flow. The critical weight has been placed at 0.01, meaning that points with a weight below 0.01 are labelled as free flow. This value is determined using Fig. 7, which shows a scatter plot of all speeds and corresponding weights of the first location. We observe that almost all low speeds (say speeds below 70 km/h), have a weight which is either zero or very close to zero. Speed-weight plots of the other four locations showed a similar pattern. Therefore, we conclude that a critical weight of 0.01 generally allows for a sensible labelling.

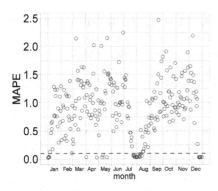

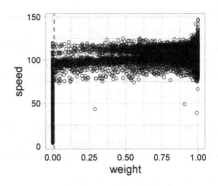

Fig. 6. Plot of the MAPE of the robust regression model for all days.

Fig. 7. Plot of the weights and corresponding speeds for location 1.

5   Conclusion and Discussion

We have developed an algorithm to identify high-performance days based on an estimation of critical speed and the breakdown probability. The algorithm is relatively straightforward and only requires two quantities; the average traffic flow and the average speed. The algorithm relies on the shape of the fundamental diagram; each observation is classified as either free flow or congestion using robust regression and the critical speed is estimated as the separating line between the two sets. Using a non-parametric estimator for the breakdown probability, we are able to quantify both characteristics of a high-performance day (roughly speaking, high speed and high flow). The algorithm has shown its capabilities by identifying high-performance days on the A15 near Papendrecht in 2018.

A natural follow-up question would be in the direction of causality. Indeed, one could wonder *why* certain days exhibit extraordinary behaviour, in terms of an unexpected absence of traffic jams. A possible explanation could be traffic homogeneity; perhaps during the high-performance days, there were fewer trucks, leading to fewer speed differences between vehicles. Alternatively, the answer may lie hidden in microscopic data; certain (desirable) behavioural characteristics of drivers might be over-represented during high-performance days. This paves the way towards reducing traffic jams from a different perspective and may lead to new insights as well as an easier investigation of countermeasures against traffic jams. This non-trivial extension is, however, beyond the scope of this paper. Instead, we present this tool to facilitate further research into countermeasures against traffic jams, as the algorithm is able to identify which days need to be studied further.

We must be critical of our approach as well, in particular in terms of generality. This mainly relates to the two (subjective) thresholds: the critical weight (to distinguish between congestion measurements and free flow measurements) and the critical level of the MAPE of the regression model (to identify a lack of congestion). Both values were determined based on the five locations of the

A15 Papendrecht 2018 data set. However, when testing the algorithm on other data sets, we still observed both a sensible labelling of the data points as well as a plausible recognition of days with little or no congestion. In fact, we tested the algorithm on data sets which violated the assumption of a constant free-flow speed and the algorithm still identified days with a high traffic flow and a striking absence of traffic jams.

Acknowledgements. This work was supported by NWO under Grant 438-13-206. We thank De Verkeersonderneming for hosting Bo Klaasse during his internship. We thank Stella Kapodistria and Onno Boxma for interesting discussions on the manuscript.

References

1. Arnesen, P., Hjelkrem, O.A.: An estimator for traffic breakdown probability based on classification of transitional breakdown events. Transp. Sci. **52**(3), 593–602 (2017). https://doi.org/10.1287/trsc.2017.0776
2. Bergendorff, P., Hearn, D.W., Ramana, M.V.: Congestion toll pricing of traffic networks. In: Pardalos, P.M., Hearn, D.W., Hager, W.W. (eds.) Network Optimization, pp. 51–71. Springer, Heidelberg (1997). https://doi.org/10.1007/978-3-642-59179-2_4
3. Bremmer, D.: Dit zijn de 20 duurste files van Nederland (2019). https://www.ad.nl/economie/dit-zijn-de-20-duurste-files-van-nederland~a4803756/. Accessed 01 Aug 2019
4. Dervisoglu, G., Gomes, G., Kwon, J., Horowitz, R., Varaiya, P.: Automatic calibration of the fundamental diagram and empirical observations on capacity. In: Transportation Research Board 88th Annual Meeting, vol. 15 (2009)
5. Ettema, D., Knockaert, J., Verhoef, E.: Using incentives as traffic management tool: empirical results of the "peak avoidance" experiment. Transp. Lett. **2**(1), 39–51 (2010). https://doi.org/10.3328/TL.2010.02.01.39-51
6. Gaddam, H.K., Rao, K.R.: Speed-density functional relationship for heterogeneous traffic data: a statistical and theoretical investigation. J. Mod. Transp. **27**(1), 61–74 (2019). https://doi.org/10.1007/s40534-018-0180-z
7. Goh, M.: Congestion management and electronic road pricing in Singapore. J. Transp. Geogr. **10**(1), 29–38 (2002). https://doi.org/10.1016/S0966-6923(01)00036-9
8. Hegyi, A., Hoogendoorn, S.P., Schreuder, M., Stoelhorst, H., Viti, F.: SPECIALIST: a dynamic speed limit control algorithm based on shock wave theory. In: 2008 11th International IEEE Conference on Intelligent Transportation Systems, pp. 827–832. IEEE (2008). https://doi.org/10.1109/ITSC.2008.4732611
9. Kerner, B.S.: Introduction to Modern Traffic Flow Theory and Control: The Long Road to Three-phase Traffic Theory. Springer, Heidelberg (2009). https://doi.org/10.1007/978-3-642-02605-8
10. Kerner, B.S., Rehborn, H.: Experimental properties of phase transitions in traffic flow. Phys. Rev. Lett. **79**(20), 4030–4033 (1997). https://doi.org/10.1103/PhysRevLett.79.4030
11. Knoop, V.L., Daamen, W.: Automatic fitting procedure for the fundamental diagram. Transp. B: Transp. Dyn. **5**(2), 129–144 (2017). https://doi.org/10.1080/21680566.2016.1256239

12. Li, H., Bertini, R.L.: Comparison of algorithms for systematic tracking of patterns of traffic congestion on freeways in Portland. Oregon. Transp. Res. Rec. **2178**(1), 101–110 (2010). https://doi.org/10.3141/2178-11
13. Montgomery, D.C., Peck, E.A., Vining, G.G.: Introduction to Linear Regression Analysis. Wiley, Hoboken (2012)
14. NDW: Home - Nationale Databank Wegverkeersgegevens (2019). https://www.ndw.nu/en/. Accessed 10 Sept 2019
15. Ong, R., et al.: Traffic jams detection using flock mining. In: Gunopulos, D., Hofmann, T., Malerba, D., Vazirgiannis, M. (eds.) ECML PKDD 2011. LNCS (LNAI), vol. 6913, pp. 650–653. Springer, Heidelberg (2011). https://doi.org/10.1007/978-3-642-23808-6_49
16. Petrovska, N., Stevanovic, A.: Traffic congestion analysis visualisation tool. In: 2015 IEEE 18th International Conference on Intelligent Transportation Systems, pp. 1489–1494. IEEE (2015). https://doi.org/10.1109/ITSC.2015.243
17. Stutz, C., Runkler, T.A.: Classification and prediction of road traffic using application-specific fuzzy clustering. IEEE Trans. Fuzzy Syst. **10**(3), 297–308 (2002). https://doi.org/10.1109/TFUZZ.2002.1006433
18. Swamidass, P.M.: MAPE (mean absolute percentage error). In: Swamidass, P.M. (ed.) Encyclopedia of Production and Manufacturing Management, pp. 462–462. Springer, Boston (2000). https://doi.org/10.1007/1-4020-0612-8_580
19. Treiber, M., Kesting, A.: Traffic Flow Dynamics: Data, Models and Simulation. Springer, Heidelberg (2013). https://doi.org/10.1007/978-3-642-32460-4
20. Tu, H.: Monitoring travel time reliability on freeways. Ph.D. thesis, TU Delft (2008)
21. Vaqar, S.A., Basir, O.: Traffic pattern detection in a partially deployed vehicular ad hoc network of vehicles. IEEE Wirel. Commun. **16**(6), 40–46 (2009). https://doi.org/10.1109/MWC.2009.5361177
22. Venables, W.N., Ripley, B.D.: Modern Applied Statistics with S-PLUS. Springer, Heidelberg (2013). https://doi.org/10.1007/978-0-387-21706-2
23. Yohai, V.J.: High breakdown-point and high efficiency robust estimates for regression. Ann. Stat. **15**(2), 642–656 (1987). https://doi.org/10.1214/aos/1176350366

A Network Aware Resource Discovery Service

Luigi Liquori[1] , Rossano Gaeta[2] , and Matteo Sereno[2(✉)]

[1] Université Côte d'Azur, INRIA Sophia Antipolis - Méditerranée,
Biot, Sophia Antipolis, France
Luigi.Liquori@inria.fr
[2] Dipartimento di Informatica, Università di Torino, Turin, Italy
{rossano.gaeta,matteo.sereno}@unito.it

Abstract. Internet in recent years has become a huge set of channels for content distribution highlighting limits and inefficiencies of the current protocol suite originally designed for host-to-host communication. In this paper we exploit recent advances in Information Centric Networks in the attempt to reshape the actual Internet infrastructure from a host-centric to a name-centric paradigm where the focus is on named data instead of machine name hosting those data. In particular, we propose a Content Name System Service that provides a new network aware Content Discovery Service. The CNS behavior and architecture uses the BGP inter-domain routing information. In particular, the service registers and discovers resource names in each Autonomous System: contents are discovered by searching through the augmented AS graph representation classifying ASes into customer, provider, and peering, as the BGP protocol does. Performance of CNS can be characterized by the fraction of Autonomous Systems that successfully locate a requested content and by the average number of CNS Servers explored during the search phase. A C-based simulator of CNS is developed and is run over real ASes topologies provided by the Center for Applied Internet Data Analysis to provide estimates of both performance indexes. Preliminary performance and sensitivity results show the CNS approach is promising and can be efficiently implemented by incrementally deploying CNS Servers.

Keywords: Discovery Service · Naming · Performance evaluation · Network and economical awareness

1 Introduction

Information Centric Networks (ICN) is a clean-state approach to redesign the actual Internet infrastructure from a host-centric, fully connected, paradigm to a name-centric, loosely connected, paradigm where the focus is on named data instead of machine name hosting those data. In the last decade many proposals raised from research to capture this new paradigm: they mainly can be grouped into two schools of thought: *Content Centric Networks* referring to the Jacobson-based vision [6,11,13], where routing is driven by fully qualified - human readable

© Springer Nature Switzerland AG 2020
M. Gribaudo et al. (Eds.): EPEW 2019, LNCS 12039, pp. 84–99, 2020.
https://doi.org/10.1007/978-3-030-44411-2_6

- hierarchical names and *Data Oriented Network Architecture* referring to a flat, unreadable but unique name-space [7] (see also [1,3,12]).

While it is always exciting to conceive a new network starting from new concepts and from a clean-state design, network's history teaches us that the Internet infrastructure and its protocol suite have little changed; this is, quite obviously, because of strong backward-compatibility needs, and because of the tremendous expansion of the Internet phenomenon.

This paper supports the evolutive research line and presents a lightweight network aware Internet Service to be implemented between the Transport and the Session layers (referring to the ISO-OSI protocol layering). We call this new service *Content Name System* (CNS) organized throughout a set of communicating CNS Servers and we design a protocol, called link, implementing the Service Discovery.

The purpose of this Discovery Service is to publish machine-IP-addresses being the owners (or the purveyors) of some named-contents and retrieve that machine-IP-addresses performing a distributed search using the named-content as the database-key. The service binds a set of IP-addresses to content-names, the latter referred by *hypernames*. The CNS Service stops when some or no IP-addresses are returned or when no other CNS Server can be delegated in the iterative call implementing the distributed data-base query. Each CNS Server is equipped with a database containing for each queried content-name, the set of corresponding IPs ordered by local awareness.

In our proposal, the CNS Servers are distributed over the Internet *Autonomous Systems* (AS): there is one CNS Server per AS (or for load balancing purposes there can be multiple CNS Servers) taking care of resources registered inside the AS itself. Furthermore, the Discovery Service leverages on AS relationships by mimicking their hierarchy; in the CNS Service each AS uses the (business) relationships with the ASes in its neighborhood to also drive the content location process according to the so called "valley-free" property, i.e., that the discover process does not generate any supplementary cost for the AS involved in the discovery. Therefore, the main contributions of the paper are:

- The definition of a Resource Discovery Service (CNS) with related protocol link allowing to search contents names through Autonomous Systems: the service is achieved by defining *(i)* a naming notation to denote contents, *(ii)* a distributed database implemented by CNS Servers deployed along ASes, and *(iii)* the link Discovery Protocol, to route queries along the ASes.
- The development of a C-based simulator of CNS and the link protocol to conduct a preliminary performance and sensitivity analysis of the proposed CNS. To this end, performance of the proposed CNS is represented by the *hit probability* (that is defined as the fraction of ASes that successfully locate a requested object) and the *average lookup length* representing the average number of CNS Servers explored during the search phase. Experiments are run over real ASes topologies provided by CAIDA [2] to provide reliable and meaningful estimates of both performance indexes; we also analyze how

high performance can be if Tier-1 ASes are excluded from the search phase providing that Tier-2 ASes are incentivized to cooperate.

The rest of the paper is organized as follows: in Sect. 2 we define the proposed CNS (we define hypernames, the CNS Servers, and the `link` protocol); Sect. 3 presents preliminary simulation results to assess the performance of CNS Service. Finally, Sect. 4 summarizes the paper contributions and discusses some further developments.

2   Content Name System

2.1   Hypernames

A *hypername* (HN) is a human readable string denoting the content name, enriched with a number of optional parameters to identify its ownership, its integrity, its hosting, and its attribute-list. Hypernames are generated by the following abstract syntax:

[`fing_princ`:] [`fing_cont`:] [`hosts`:] [`tags`:] `cont_name`, where

- `cont_name` is a (possibly human readable) string denoting a content name (e.g. "openoffice.iso", "traffic_light", "defibrillator", "plastic_bottle", "pedestrian", URI, MAC, GUID, etc.);
- `tags` is an optional (possibly human readable) list of keywords (e.g. "sell", "buy", "rent", "cars", etc) associated with a given content;
- `hosts` is an optional list of hostnames being the purveyors of the content: when a hypername contains a list of hostnames, then the content name is retrieved from one of the hostnames: the local CNS perform a DNS query, transforms one (or all) hostname(s) into IP address(es) and return that list to the sender of the discovery request;
- `fing_cont` is an optional digital signature (hash) denoting the integrity of the content to be retrieved;
- `fing_princ` is an optional digital signature denoting the public asymmetric key of the principal, i.e. the owner of the content: it allow to identify the identity of the latter as soon as we retrieve the content itself.

Therefore, a hypername is characterized by a human readable part and another part which is human unreadable.

2.2   CNS Servers

Similarly to the well-known DNS Service, we locate the CNS Server in the ISO-OSI hourglass at the application level. The goal of CNS is to translate hypernames into lists of IP addresses, that is $HN \implies \{IP_i\}^{i \in I}$ with $I = \emptyset$ in case of discovery failure.

BGP Business Relationship Among ASes. CNS servers are distributed over the ASes; more precisely, there is one CNS per AS (or for load balancing purposes there can be multiple servers) taking care of resources registered inside the AS itself. The Discovery Service leverages on AS relationships: the CNS server hierarchy mimics the AS relationship hierarchy. It is well-known that the routing between ASs (also called interdomain routing) is determined by the *Border Gateway Protocol* (BGP) [10]. The main feature of the interdomain routing is that it allows each AS to choose its own administrative policy in selecting the best route, and announcing and accepting routes.

The commercial agreements between two ASes domains can be classified into two main classes of agreements: *customer to provider*, and *peering*[1]. An AS customer pays its AS provider (or ASes in case of multiple providers) for connectivity to the rest of the Internet. A pair of ASes can set up a peering relation and in this case they agree to exchange traffic between their respective customers free of charge.

The AS-graph annotated with these two kinds of relationships is one of the most famous and studied representations of the Internet (e.g., see the measurement studies provided by CAIDA [2].) In this graph, according to their roles, we can distinguish three different kind of ASes: Tier-1, Tier-2, and Tier-3. A Tier-1 is an AS that can reach every other destination on the Internet without paying other ASes. In other words, a Tier-1 is an AS with (many) customer ASes but with no provider. For connectivity purposes, the Tier-1-s set up peering relationships among them. On the other hand, a Tier-3 is a stub AS, without any transit customers, and with some peering relationships. Tier-3 ASes generally purchase transit Internet connection from Tier-2 ASes and, in some cases, even from the Tier-1 ASes as well. Finally, a Tier-2 is an AS with customers, and some peering, but that still buys transit service from Tier-1 ASes to reach some portion of the Internet.

The relationships among the ASes play a fundamental role in shaping the AS graph structure and in defining the routing policies implemented throughout BGP. In particular, the paths between two ASes must avoid routing policies that would result in unjustified payments by some AS. Examples of such incorrect routing paths are, for instance, an AS provider that routes the traffic directed to another AS provider by forwarding it to one of its AS customers. This path is incorrect because would cause an unjustified cost in charge to the AS customer used as an intermediate. Another example of incorrect path occurs when an AS forwards its traffic by using as intermediate step one of its peering relationships. In this case the peer AS chosen as intermediate would be in charge of the transit cost for the traffic it forwards. Figure 1 (presented in [5]) shows a simple configuration of seven ASes, their relationships (i.e., provider-to-customer and peering). On this simple AS graph we report two wrong, and two correct paths.

The routing paths among pairs of ASes is obtained by the BGP protocol that uses selective exporting path rules (i.e., each AS selectively provides transit

---

[1] The ASes can establish also other type of relationships such as "sibling" and "backup". For the purposes of this paper we neglect them.

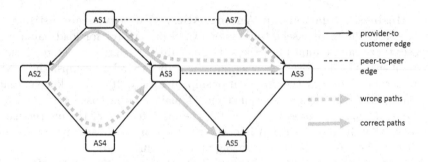

Fig. 1. A simple AS graph with two wrong and two correct routing paths.

services for its neighboring ASes). All the paths (also called routes) with property we previously discussed are called *free-valley* (or *no-valley*) paths [5].

CNS Hierarchical Topology. The CNS distributed database is organized into a hierarchy of CNS servers deployed according to the Tier-1/Tier-2/Tier-3 AS topology. Each AS must have at least one CNS, called *authoritative*, whose database will take into account the association of each hypername with a list of IPs that have registered a content named by a hypername. The authoritative CNS also knows exactly its position in the distributed database, namely *(i)* the IP addresses of all customers' CNS, *(ii)* the IP addresses of all providers' CNS and *(iii)* the IP addresses of all peer-to-peer CNSs: this will allow to dispatch queries along the distributed database.

In order to make a content discoverable, the owner or purveyor publishes an hypername referring to a content in the local CNS. Note that the publication in a CNS associates the hypername with a principal, and that principal holds the content as an owner or a purveyor (the content being mutable or immutable). Suppose a given content be available by a host belonging to an autonomous system: the host can publish, through the CNS Service, the content in the authoritative CNS local database. To do this, at the beginning, the host creates a proper hypername that will be sent as a formal parameter to the authoritative CNS. Note that the host decides which attribute to attach to the hypername and if it should publish that content as a owner or as a purveyor. In the first case (owner) the publication is done by a simple write in the CNS' database[2]. In the second case (purveyor) the host could be asked to package a `.torrent` file and write it in the CNS' database. Following the Bittorrent jargon, the purveyor plays a role of seed and it will be asked to publish itself as a purveyor of the content every time interval: further nodes entering the swarm for the content will be asked to publish his name in the torrent; for that content, the CNS server would serve as a kind of network aware Bittorrent tracker.

---

[2] Depending on a local policy, the CNS could ask to republish the content every n seconds.

```
1.01 on receipt of link(HN,DOWN) from provider do  // receive a query from a "downhill"
1.02 value = lookupdb(HN);  // search HN in the CNS' local data base
1.03 if (value ≠ 0) // some IP publishing HN are found
1.04 then {publish(HN,value) to CNS; return value to provider};  // write in the local CNS
              and return IPs "back to the downhill"
1.05 else list = select(α,customerlist);  // select some customers CNS
1.06 forall cus ∈ list do value = value ∪ send link(HN,DOWN) to cus;  // and forward   the
              query downhill through a customer
1.07 publish(HN,value) to CNS; return value to provider;  // write in the local CNS and
              return IPs "back to the hill"
```

Fig. 2. Queries from provider with downhill direction continue on α thread downhill.

```
2.01 on receipt of link(HN,UP) from peer do  // receive a query from a peer on the "top of the hill"
2.02 value = lookupdb(HN);  // search HN in the CNS' local data base
2.03 if (value ≠ 0) // some IP publishing HN are found
2.04 then {publish(HN,value) to CNS; return value to peer};  // write in the local CNS and
              return IPs "back to the top of the hill"
2.05 else list = select(α,customerlist);  // select some customers CNS
2.06 forall cus ∈ list do value = value ∪ send link(HN,DOWN) to cus;  // and forward   the
              query but downhill through a customer
2.07 publish(HN,value) to CNS; return value to peer;  // write in the local CNS and
              return IPs "back to the top of the hill"
```

Fig. 3. Queries from peer with uphill direction will change on α thread downhill.

2.3   The link Discovery Protocol

Each autonomous system holds an authoritative CNS server, that records the mappings for all the hypernames published inside it. The CNS database is organized hierarchically following CAIDA's augmented graph [2]. Let α, β and γ being AS-specific parameters; in a nutshell, the link protocol proceeds intuitively as follows:

1. the client first contacts its authoritative CNS and then searches the hypername in the local publications (i.e. in the current and in the peering CNS);
2. if the above fail, then the authoritative CNS forwards the query through α-CNS belonging to ASes in "downstream", i.e., with which we have signed some provider-to-customer agreement;
3. if the above fails, then the authoritative CNS forward the query through γ-CNS belonging to ASes in peer, i.e., with which we have signed some peering-to-peering agreement;
4. if all of the the above fail, then the authoritative CNS forward the query through β-CNS belonging to ASes in "upstream", i.e., with which we have signed some customer-to-provider agreement.

The link pseudocode is presented in Figs. 2, 3, and 4. A client sends a query to the local authoritative CNS server, with argument the hypername HN and a direction UP (from customer-to-provider or from peer-to-peer) or DOWN (from provider-to-customer). This query is recursive and the client will be blocked until the CNS will answer positively with a result containing a set of addresses $\{IP_i\}^{i \in I}$ associated with HN, or with a search failure.

```
3.01 on receipt of link(HN,UP) from customer do  // receive a query from a "uphill"
3.02 value = lookupdb(HN);  // search HN in the CNS' local data base
3.03 if (value ≠ 0)  // some IP publishing HN are found
3.04 then {publish(HN,value) to CNS; return value to customer};  // write in the local CNS and
            return IPs "back to the uphill"
3.05 else list = select(α,customerlist);  // select some customers CNS
3.06 forall cus ∈ list do value = value ∪ send link(HN,DOWN) to cus;  // and forward the
            query but downhill through a customer
3.07 if (value ≠ 0)  // some CNS are suggested
3.08 then {publish(HN,value) to CNS; return value to customer};  // write in the local CNS and
            return IPs "back to the uphill"
3.09 else list = select(γ,peerlist);  // select some peers CNS
3.10 forall per ∈ list do value = value ∪ send link(HN,UP) to per;  // and forward the query
            uphill through a top of the hill peer
3.11 if (value ≠ 0)  // some CNS are suggested
3.12 then {publish(HN,value) to CNS; return value to customer};  // write in the local CNS
            and return IPs "back to the uphill"
3.13 else list = select(β,providerlist);  // select some provider CNS
3.14 forall pro ∈ list do value = value ∪ send link(HN,UP) to pro;  // and forward the query
            uphill through a provider
3.15 publish(HN,value) to CNS; return value to customer  // write in the local CNS and return
            IPs "back to the uphill"
```

Fig. 4. A query from customer with uphill direction will continue on three directions: first α-downhill, then γ-downhill, and finally β-uphill.

Location Process Start. A client sends a query to the authoritative CNS server where the client belongs to, with argument the hypername HN. In DNS jargon, this query is recursive i.e., the client will be blocked until the CNS will answer positively with a result containing a set of addresses $\{IP_i\}^{i \in I}$ associated with HN, or with a search failure.

Figure 2: from Provider with Downhill Direction. This code refers to the general case when the current CNS receives a link message with a HN and a downhill direction from a provider-CNS (line 1.01). First of all, a local lookup is performed (1.02); in case of success, the result value is returned to the sender[3] (1.04); else selects α-customer-CNS (1.05) and sends α-iterative link queries with the same HN and the same downhill direction (1.06); then collects the result value and send it back to the sender of the first link message (1.07). Before return, all the results will be written in the local CNS in order to give a direct answer in successive queries.

Figure 3: from Peer with Uphill Direction. Following the BGP jargon, this code refers to the case of being "on the top of the hill", i.e., receiving a message from uphill and from a peer-to-peer-CNS. Execute the same code as the one of Fig. 2, with the following exception: invert the direction from uphill to downhill when sending α-*iterative* link queries (2.06). Before return, all the results will be written in the local CNS in order to give a direct answer in successive queries.

---

[3] At the beginning of the search, the sender is just the authoritative-CNS itself, while in the middle of the location process, the sender is a provider-CNS.

Figure 4: from Customer with Uphill Direction. This code refers to the case where a CNS receives a `link` message from uphill from a customer-CNS. Following the BGP jargon, when we receive a query from a customer and with an uphill direction the following steps are executed. First of all, a local lookup is performed (line 3.02): in case of success, the result value is returned to the sender (3.04); else select α-customer-CNS[4] (3.05) and send α-iterative `link` queries with the same HN but inverting the direction from uphill to downhill (push downhill the query) (3.06); in case of success, the result value is returned to the sender (3.08); else select γ-peer-CNS (3.09) and send γ-iterative `link` queries with the same HN and the same direction[5] (3.10); in case of success, the result value is returned to the sender (3.12); else select β-provider-CNS (3.13) and send β-iterative `link` queries with the same hypername and the same uphill direction (in other words: go uphill only after tried to invert the search downhill but all the queries failed) (3.14); as the last resort of the query, return a success or failure value to the sender. As in [6], before return all the results will be written in the local CNS in order to give a direct answer in successive queries.

Note. All α, β, and γ are dependent on the local CNS; all messages not matching with the above pseudocode are flushed by the receiving CNS server.

2.4 CNS vs DNS

Because of its resemblance with the DNS Service, we highlight differences and similarities in the following:

- DNS [9] is a fundamental phone book directory for the Internet. It mainly uses the UDP transport to query other distributed DNS servers to answer client questions like "which IP addresses are associated with the name www.google.com?" The DNS Service provides information about hosts querying the DNS hierarchy: this hierarchy can go through ASes and does not follow the AS cash flow route: the small amount of packets involved in DNS resolution makes DNS economically scalable. On the contrary, and this has been made explicit in the `link` pseudocode, packets will be routed following the economic interest of the AS that generates the query: this point is crucial for ensuring the Discovery Service to be economically scalable.
- DNS delegates name resolution into domain zones from the smallest to the biggest zone. With the same idea, the CNS delegates content discovery (content name resolution) through ASes always trying to follow, when possible, a reverse cash flow route in order to suggest to the further content delivery an ordinary cash flow route;
- DNS distributed database is indexed via domain names. On the other hand, the relations among CNS servers are derived by the relations among ASes (customer-to-provider, provider-to-customer and peering relations). These

---

[4] Do not choose the customer-CNS that have sent the query.

[5] Successive execution of code in Fig. 3 will later invert the direction from uphill to downhill, i.e. we push downhill the query.

relations can be derived by using CAIDA's AS relationships dataset maps (see [2] and Gao's pioneering work [5] on valley-free routing);
– DNS queries can be iterative or recursive: the same holds for CNS: nevertheless, an efficient implementation of the CNS Service prefers iterative queries.

3   Performance Results

In this section we present results that characterise the performance of the CNS Service and of the `link` protocol. Simulations show that link is able to successfully locate objects with high probability at low cost; they also show that good performance can be obtained by excluding Tier-1 ASes from the search phase providing that Tier-2 ASes are incentivized to cooperate.

Performance is represented by two indexes: the *hit probability* (denoted as p_{hit}) that is defined as the fraction of ASes that successfully locate a requested object, and the *average lookup length* (denoted as avg_{ll}) representing the average number of CNS servers explored during the search phase. To this end, we developed a C-based simulator of the proposed object Discovery Service. The simulator runs by using real ASes topologies provided by CAIDA [2] and is able to reproduce the dynamic behavior of location requests. We provide a sensitivity analysis of the lookup algorithm `link` with respect to parameters α, β, and γ. We also discuss the performance of `link` as a function of the fraction of ASes that actually deploy a CNS server to support the location service.

3.1   Scenario

In our experiments we selected an ASes topology provided by CAIDA containing all the ASes and their type of relationships. In these snapshots edges between two nodes either represent peer relationships between ASes (undirected edges) or provider-to-consumer roles (directed edges). We classify ASes in Tier-1/Tier-2/Tier-3 [2] subsets based on the topological characteristics of nodes. In particular, we define as: *(i)* Tier-1 those snapshot nodes with no incoming edges, i.e., ASes that have no providers; *(ii)* Tier-3 those snapshot nodes with no outgoing edges, i.e., ASes that have no customers; *(iii)* Tier-2 all other snapshot nodes. We consider a resource whose popularity is equal to 0.1 among Tier-2 and Tier-3 ASes; we assume Tier-1 ASes do not hold a copy of the resource. Furthermore, we run the simulator by restricting location requests to only Tier-2 and Tier-3 ASes. To summarize, Tier-1 ASes participate in the search process but do not contribute any further.

3.2   Sensitivity to Lookup Parameters

We characterize the performance of `link` by relying only on customers, i.e., parameters β and γ are both equal to 0. In this case, Tier-3 ASes can successfully resolve the location request only if they hold a copy of the resource. Tier-2 ASes can exploit more search path, instead. Indeed, their p_{hit} is higher than the

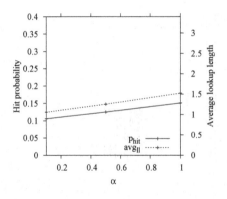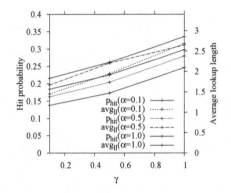

Fig. 5. Values of p_{hit} and avg_{ll} during the lookup (left plot as function of α with $\beta = \gamma = 0$, right plot as function of γ with $\beta = 0$).

resource popularity and the overall results are represented in Fig. 5 (left plot). It can be noted that increasing α raises p_{hit} from 0.105902 to 0.151068: the performance for $\alpha = 1$ represents an upper bound on the achievable performance. Furthermore, the rather small values of avg_{ll} for all considered values of α show that a little number of search requests contacts more than one CNS.

To evaluate the impact peers in the AS snapshot we consider all combinations of parameters α and γ where $\beta = 0$ in results we present in Fig. 5 (right plot). It can be noted that parameter γ has moderate impact since Tier-3 ASes have very limited peer AS relationships. On the contrary, Tier-2 ASes can exploit their peering relations to increase their p_{hit} although the maximum achievable performance is just 0.298151.

The impact of parameter β on the performance of link is remarkable, instead. It can be noted from results in Fig. 6 that by increasing β from 0.1 to 0.5 for a very low value of α (0.1) we obtain $p_{hit} = 0.562925$ (from the value 0.107914). By further increasing it to 1 we obtain that location requests are successfully served almost surely for any value of α. Of course, this improvement is paid by the increased cost of the service in terms of the avg_{ll} values.

The last set of results we present is to analyze how the resource popularity impacts on the cost of lookups and how effectively link is able to successfully serve location requests. To this end, we considered the triple of parameters $(\alpha, \beta, \gamma) = (0.1, 0.75, 0.1)$ and performed location requests for increasingly rare objects. We chose these low values for link parameters because it aims at avoiding that the search phase (and as a byproduct the resource exchange) indiscriminately jumps on the different network locations thus possibly increasing transit fees.

Figure 7 shows results that link yields values of p_{hit} that are order of magnitudes higher than resource popularity even for rather scarce object diffusion. Of course, the scarcer the resource the higher the number of CNS to contact before finding one that owns a copy. Indeed, avg_{ll} values increase as resource

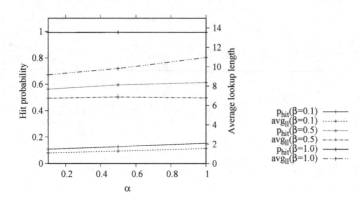

Fig. 6. Values of p_{hit} and avg_{ll} during the lookup as function of pair (α, β) for $\gamma = 0$.

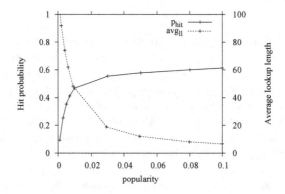

Fig. 7. Values of p_{hit} and avg_{ll} during the lookup as a function of popularity for parameters $(\alpha, \beta, \gamma) = (0.1, 0.75, 0.1)$.

popularity decreases although the average lookup length for the scarcer resource is only 0.2% of the size of the AS snapshot we use for experiments.

As a final remark, please note that although the analysis we presented does not account for the dynamic evolution of the resource popularity (i.e., we are assuming here that the resource popularity does not change during the lookup phase), the insight it provides can be used by the CNSs to explore a wide set of parameters vs. the resource popularity.

In particular, performance can be tuned by letting each CNS modulate the costs of the lookup phase in terms of number of explored CNSs (and hence of the distance in terms of AS hops). In other words, the lookup algorithm can modulate the CNS's network awareness by tuning parameters (α, β, γ) to balance costs and expected p_{hit} since each CNS is aware of its connectivity relations and of the transit costs related with these relations.

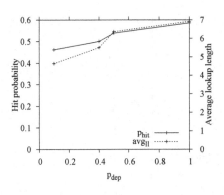 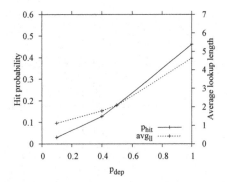

Fig. 8. Values of p_{hit} and avg_{ll} during the lookup as a function of p_{dep} for parameters $(\alpha, \beta, \gamma) = (0.1, 0.75, 0.1)$ and popularity 0.1 (left). The case with $(\alpha, \beta, \gamma) = (0.1, 0.75, 0.1)$ and same resource popularity for un-cooperating Tier-1 ASes (right).

3.3   Sensitivity to Deployment of CNS

Here we evaluate the performance of link as a function of how widespread CNS are in the entire network.

To this end, Fig. 8 (left plot) shows results when Tier-1 ASes deploy a CNS with a certain probability p_{dep} for $(\alpha, \beta, \gamma) = (0.1, 0.75, 0.1)$ and resource popularity equal to 0.1.

It can be noted the contribution of Tier-1 ASes to the performance of link is not so high. Indeed, when Tier-1 ASes do not cooperate during the lookup we obtain $p_{hit} = 0.461884$ that is moderately less than the highest possible value, i.e., 0.58548. This can be explained by noting that Tier-2 ASes are generally well connected with many peers and many customers. This means that link can easily give up Tier-1 ASes and still be able to provide very good chances to successfully locate objects.

We further consider the case where no Tier-1 AS deploys a CNS and both Tier-2 and Tier-3 cooperate with probability p_{dep}. Results are summarized in Fig. 8 (right plot); it can be noted that at least 40% of ASes should deploy a CNS to obtain a hit probability value that is greater than the resource popularity.

The last set of results characterizes a system where Tier-1 ASes do not cooperate while all Tier-2 ASes do. We consider varying levels of cooperation of Tier-3 ASes (the majority of ASes in the CAIDA snapshot) modeled by the adoption probability p_{dep}.

Results are reported in Fig. 9; they show that link performance are only slightly degraded when only 10% of Tier-3 ASes cooperate in the search process by adopting a CNS. This means that adoption of a CNS can progressively start by incentivizing Tier-2 ASes to participate in the lookup framework.

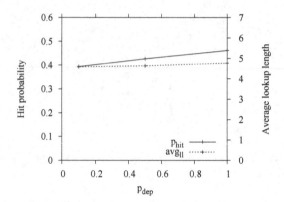

Fig. 9. Values of p_{hit} and avg_{ll} during the lookup for cooperating Tier-2 ASes, as a function of cooperation of Tier-3 ASes (probability p_{dep}) for parameters $(\alpha, \beta, \gamma) = (0.1, 0.75, 0.1)$ and popularity 0.1.

4   Further Developments

This section presents some improvements and features that could be explored and included in CNS Service.

Discovery Improvements

1. To improve queries hit and limit messages, CNS can put in a cache the result of a successful queries lookup giving positive results not in the current AS. The positive effect of caches applied to all the CNS databases can leverage the number of message exchanges between CNSes;
2. To reduce traffic and flooding attacks, each CNS can limit the number of link packets arriving from an AS-customer, AS-provider and AS-peer; their number can be fixed on a AS-to-AS basis;
3. To improve liveness, a liveness politics can be implemented (see the Kademlia's bucket-table ordering republication [8]). Each publication in an authoritative CNS can have a lifespan: after the end of the lifespan, either the publisher re-publish the content in the CNS, or the record is simply dropped out from the CNS;
4. To limit the research space, a TTL can be introduced in link messages; TTL allows to limits the lifetime of lookup messages. A TTL counter attached to each link message allows to flush messages whose counter has elapsed;
5. To improve participation, incentives to locally republish contents retrieved abroad can be introduced: republication can be a simple pointer to another CNS. A tit-for-tat strategy could be installed between clients (looking for contents) and purveyors (distributing the contents) were the CNS should play a special role being in the middle of the above two actors;
6. To improve load distribution, CNS can perform load distribution among replicated copies of a single content. If CNS tables map a hypername into a lists

of IP, then the CNS can respond with the entire list of purveyors, or it can rotate the ordering of the addresses within each reply. As such, IP rotation performed by CNS can distribute among multiple purveyors;

7. To improve the discovery success rate and focus the discovery search, each CNS can dynamically refine their α, β and γ flooding parameters by combining with the success probability of a given tag in the previous queries.

Content Aggregation in CNS. The data quality can be compromised by many factors, including data entry errors (*"OpneOffice"* instead of *"OpenOffice"*), missing integrity constraints (*"eat before December 12018"*), multiple convention (*"1ˢᵗ, rue Prés. Wilson, Antibes"*, versus *"1, rue du Président Wilson, Antibes"*), optional arguments ("+33(0)678123456" versus "0033678123456"), see [4] for a survey of data deduplication techniques. For a simple intuition, let the following hypername:

HN1 = fing_cont:hosts1:tags1:cont_name1

be published in some CNS and let

HN2 = fing_cont:hosts2:tags2:cont_name2

be retrieved by a `link` query: HN1 and HN2 differ in content names and in all logical attributes but the digital signature of the content `fing_cont`, which is the same. Because the digital signature is the same, the two hypernames should be merged into a single one. More generally, each time a purveyor publishes an immutable content with a given HN2, or a query return a list of purveyors, the authoritative CNS should verify that the same content is not already published with a similar but equationally different HN1[6] and, when it is the case, merge the two entries. Content aggregation should rewrite the previous two entries and substitute with the following ones:

HN1 = link to HN3
HN2 = link to HN3
HN3 = fing_cont:hosts1,hosts2:tags1,tags2:cont_name1|cont_name2

where the symbol "," denotes list concatenation and the symbol "|" denotes an "or" operator that allow to match both content names in pattern matching.

Mobility. Since traffic from wireless and mobile devices has exceeded traffic from wired devices, most contents are requested and delivered by both wireless and mobile devices. It is well known that wireless and mobile devices may easily switch networks, changing their IP address and thus introducing new communication modalities based on intermittent and, possibly, opportunistic connectivity [12]. The CNS Service Discovery should be able to deal with mobility in case the owner/purveyor is a mobile host.

---

[6] E.g. synchronizing mail or telephone contact across multiple google accounts.

Nomadism. When a mobile node wants to publish a content, two cases can happen according to the (im)mutability of the content:

- *immutable:* (most common of the two). The authoritative CNS related to the mobile Internet provider accept the publication of an immutable content by a mobile user with the proviso of *(i)* recording the identity of the user, via e.g. the MAC address of the mobile device (or another identifier of the mobile node), and *(ii)* asking to the mobile user to re-publish the content more frequently than a fixed device, and *(iii)* possibly blacklisting a mobile device that appear and disappear too fast or too often.
- *mutable:* it deal with the possibility to keep an identity also in case the user is navigating through different mobile networks. The authoritative CNS related to the mobile ISP could accept the publication of a mutable content if and only if the logical attribute `fing_princ` is present and the logical attribute `hosts` contains only one symbolic name or only one IP.

Security. Until now, a few significant DNS attacks has corrupted the DNS service: this is because *(i)* DNS Servers are machines managed and protected by system administrators, *(ii)* the DNS protocol pushes lookup always "below" the hierarchical database, minimizing the "uphill ascents", and *(iii)* of making use of cache techniques. We think that the above arguments could be applied also the CNS Service because the relatively fixed number of CNS servers ($\sim$70K) could be managed by AS system administrators, and `link` always pushes the location process first downhill the customer-CNS distributed database, and, only in case of failure, uphill through a peer-CNS or a provider-CNS. Nevertheless, the CNS Discovery Service is not vaccinated by either DDoS bandwidth-flooding attack, or man in the middle attack, or poisoning attack, or spoofing an IP of a node below an authoritative CNS.

Acknowledgments. The work has been partially supported by the HOME (Hierarchical Open Manufacturing Europe) project, supported by the Regione Piemonte, Italia (framework program POR FESR 14/20).

References

1. Bari, M.F., et al.: A survey of naming and routing in information-centric networks. IEEE Commun. Mag. **50**(12), 44–53 (2012)
2. CAIDA: Center for Applied Internet Data Analysis: AS relationship (2016). http://www.caida.org/data/as-relationships/
3. Chand, R., Cosnard, M., Liquori, L.: Powerful resource discovery for arigatoni overlay network. Future Gener. Comput. Syst. **24**(1), 31–48 (2008)
4. Elmagarmid, A.K., Ipeirotis, P.G., Verykios, V.S.: Duplicate record detection: a survey. IEEE Trans. Knowl. Data Eng. **19**(1), 1–16 (2007)
5. Gao, L.: On inferring autonomous system relationships in the internet. IEEE/ACM Trans. Netw. **9**(6), 733–745 (2001)
6. Jacobson, V., et al.: Networking named content. In: Proceedings of CoNEXT. ACM (2009)

7. Koponen, T., et al.: A data-oriented (and beyond) network architecture. SIG-COMM Comput. Commun. Rev. **37**(4), 181–192 (2007)
8. Maymounkov, P., Mazières, D.: Kademlia: a peer-to-peer information system based on the XOR metric. In: Druschel, P., Kaashoek, F., Rowstron, A. (eds.) IPTPS 2002. LNCS, vol. 2429, pp. 53–65. Springer, Heidelberg (2002). https://doi.org/10.1007/3-540-45748-8_5
9. Mockapetris, P.: Domain names - concepts and facilities (1983). https://tools.ietf.org/html/rfc882. RCF 883, updated 973, 1034, 1035
10. Rekhter, Y., Li, T.: A Border Gateway Protocol 4 (BGP-4) (1995). https://tools.ietf.org/html/rfc4271. RCF 4271, obsoletes 1654, 1267, 1163, 1105
11. Shang, W., et al.: Named data networking of things. In: Proceedings of IEEE IoTDI (2016)
12. Xylomenos, G., et al.: A survey of information-centric networking research. IEEE Commun. Surv. Tutor. **16**(2), 1024–1049 (2014)
13. Zhang, L., et al.: Named data networking. Comput. Commun. Rev. **44**(3), 66–73 (2014)

EthExplorer: A Tool for Forensic Analysis of the Ethereum Blockchain

Yuriy Marchenko[1] (ID), William J. Knottenbelt[2] (ID), and Katinka Wolter[1](✉) (ID)

[1] Free University Berlin, Takustr. 9, 14195 Berlin, Germany
yuriy.marchenko@gmail.com, katinka.wolter@fu-berlin.de
[2] Imperial College London, London SW7 2AZ, UK
wjk@imperial.ac.uk

Abstract. This paper presents EthExplorer, a graph-based tool for analysing the Ethereum blockchain. EthExplorer has been designed for the assessment of Ethereum transactions, which represent diverse and complex activities in a large-scale distributed system. EthExplorer shows Ethereum addresses as nodes and transactions as directed arcs between addresses. The graph is annotated in several ways: arcs are scaled according to the amount of Ether they carry and the nodes are colour encoded to indicate types of addresses, such as exchanges, miners or mining pools. Ether transfer transactions and smart contracts are distinguished by line styles. EthExplorer can be used to trace the flow of Ether between addresses. For a given address all its output or input transactions with the corresponding receiver or sender addresses can be found. The set of considered addresses can be increased by adding selected addresses to the set of analysed addresses.

Keywords: Ethereum · Blockchain · Graphical analysis

1   Introduction

Blockchain technologies and cryptocurrencies have received much attention in recent years. Ethereum [1,2] is to date the second largest cryptocurrency after Bitcoin [3] in terms of market valuation. Ethereum allows not only for the exchange of Ether between nodes, but also hosts programs (smart contracts) [1], as highly complex transactions. Smart contracts can be used to define new *tokens* on top of the Ethereum blockchain.

To date, the Ethereum blockchain consists of 8.6 million blocks, each holding on average approximately 130 transactions. Hence, there are many transactions on the Ethereum blockchain and even though the blockchain is public finding relevant information is challenging. Graph theoretic analysis of the Ethereum blockchain has been performed in recent years [4–6].

A number of blockchain explorers exist: an live graphical presentation of the Bitcoin blockchain has been developed for the data observatory at the Data Science Institute in London [7], https://etherscan.io/ allows to retrieve information on Ethereum transactions, addresses and blocks, https://ethplorer.io/ is

© Springer Nature Switzerland AG 2020
M. Gribaudo et al. (Eds.): EPEW 2019, LNCS 12039, pp. 100–117, 2020.
https://doi.org/10.1007/978-3-030-44411-2_7

specialised on Ethereum tokens and displays additional token market information. A number of other text based information sites for the Ethereum blockchain exist (https://www.etherchain.org/, https://blockscout.com and https://enjinx. io/ are non-exhaustive examples from this list). https://ethtective.com/ is similar to our EthExplorer in that it shows addresses as nodes in a graph, transaction as links between the nodes. It has some scaling according to amount, but much less annotation and selection features than EthExplorer.

EthViewer (http://ethviewer.live/) provides an insightful real-time graphical view of the mechanics of Ethereum by showing the generated individual transactions and how they are gathered into the next mined block, which is then appended to the chain of blocks. Valid blocks are shown in green, while uncle blocks are red. The tool has great educational value, but does not allow to trace the flow of Ether.

EthExplorer is different from the above in that it facilitates understanding the flow of Ether between the Ethereum addresses, rather than explaining the mechanics of Ethereum. Therefore, EthExplorer uses a database created from the Ethereum blockchain which has been enriched with additional information, such as the types of addresses and transactions, as well as the names connected to Ethereum addresses that can be found on etherscan [8]. The annotation process makes use of heuristic arguments on the large, complex, distributed Ethereum system. The display of addresses and transactions uses colour encoding to denote the different types. Selection features allow to aggregate transactions to provide a broad overview of the activities of one or more addresses as well as to dig into the details of an addresses Ethereum business. EthExplorer is made for the qualitative and quantitative assessment of Ethereum activities, a large-scale secure distributed system. In experimental exemplary studies we aim at illustrating the potential and limits of the confidentiality of the Ethereum system. The paper is organised as follows. We first briefly introduce the fundamentals of blockchain technologies and the Ethereum blockchain in the next two sections. Then we introduce our tool EthExplorer, its software architecture and the database design. In Sect. 5 we will discuss observations we have made for the Ethereum blockchain and in Sect. 4.2 we illustrate how to use EthExplorer in a case study and show what insights can be gained using the tool. We will also discuss the limitations of this type of analysis. Section 6 concludes the paper.

2   Blockchain Basics

A blockchain as used by Bitcoin or Ethereum can be seen as a chain of blocks that is held in redundant copies by a P2P network of *full nodes*. A block refers to the previous one by including a hash pointer as reference. Blocks are created by *miners*. A miner of a proof-of-work (PoW) blockchain collects a number of transactions to be included in the block, generates the block header and solves a hash puzzle. The challenge in solving the hash puzzle is to find a *nonce* such that the hash of the block header including the nonce will stay below the *target*. The target is adjusted in regular intervals as to keep the mean time between subsequent blocks found by the network of miners constant.

A full node can create a *mining pool* by delegating sub-problems of the hash puzzle to a group of individual miners. In the reverse perspective, miners join forces in a mining pool. The mining pool behaves like a miner with a lot of hash power, since solving the hash puzzle essentially requires performing a large number of hash operations in the attempt to find a valid nonce. The mining pool will then distribute most of the reward it obtains for finding a valid block to the miners according to some strategy [9].

In Bitcoin or Ethereum the respective currency can be obtained either as mining reward, from transaction fees of transactions included in the generated block, or by exchanging it for a fiat currency (or another cryptocurrency) at an exchange. Those exchanges are similar to a stock exchange in that they typically facilitate the exchange of units of cryptocurrency for units of fiat currency. In consequence, exchanges are places where the familiar currencies of the physical world meet the more esoteric cryptocurrencies of the virtual world.

3   Ethereum

Ethereum is more than a pure blockchain; it rather is a platform for cryptoassets. Its potential beyond a cryptocurrency like Bitcoin is primarily due to the programs (called *smart contracts*) written in the Turing-complete language *Solidity*. Smart contracts can define new currencies in the form of (*tokens*) on top of Ethereum or deliver other services. A token in essence is a smart contract that follows a particular standard. Currently there are ten different token standards, which are denoted by their ERC (Ethereum Request for Comment) number. The most common tokens are the ERC-20 [10] and ERC-721 [11] tokens. Smart contracts are executed on the Ethereum platform when they are fuelled by *gas* in a transaction.

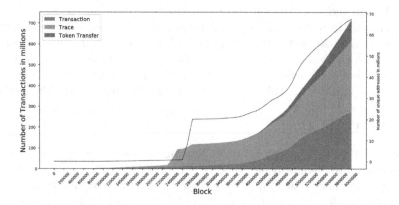

Fig. 1. Ethereum addresses and transactions (Color figure online)

The Ethereum blockchain is a PoW blockchain as described above. It was first released in the summer of 2015. Addresses in Ethereum are called *accounts* and Ethereum uses two types of accounts: *externally owned accounts (EOAs)* and *contract accounts*. Both types of accounts can hold an Ether balance. Externally owned accounts are similar to addresses in Bitcoin; they can be used as sender and receiver account in transactions to send or receive Ether in interaction with other accounts. Contract accounts have an attached program byte code (the smart contract) which can be executed when used in a transaction. We will only distinguish between the two types of accounts where necessary and will call accounts also addresses.

Transactions triggering a smart contract must be equipped with some amount of Ether, the *gas,* to limit the runtime of the smart contract, while a transaction fee is paid to the miner who includes the transaction in its block. This payment incentivises miners to perform the necessary work on a transaction and it limits the runtime of a smart contract, hence avoiding infinitely running programs.

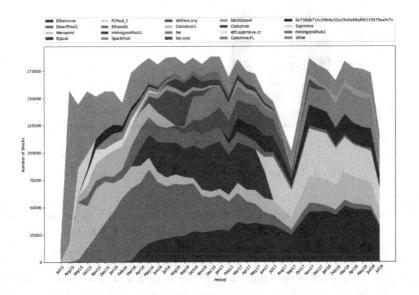

Fig. 2. Number of Ethereum blocks found

Figure 1 shows some statistics on the Ethereum blockchain collected at the end of 2018, when it had approximately 700 million transactions made by 65 million addresses (shown as the blue line with respect to the right y-axis). The graph shows the increasing number of traces over time. Traces are smart contract interactions (such as function calls, or token transfers), or internal transactions that are not permanently kept on the blockchain. The graph has been created using the data stored in the database as described in Sect. 4.

The role of mining pools is illustrated in Fig. 2, which analyses the blocks found over time between July 2015 and July 2018. While the difficulty target is set as to keep the blocks found over time constant, there is still some fluctuation and in particular in September 2017 there was a crash in the mining activity in Ethereum, following a value crash in June of the same year.

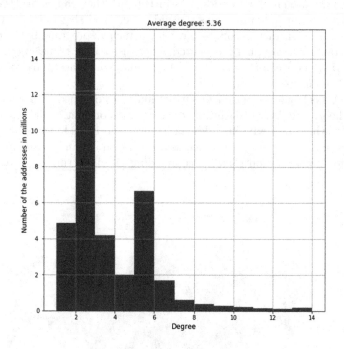

Fig. 3. Degree distribution of the Ethereum nodes

The figure also shows that the 12 largest miners (mostly mining pools) found 90% of all blocks. This demonstrates the dominance of the mining pools in Ethereum mining.

Compared to the number of addresses in Ethereum, only very few have a high degree of activity with other addresses, as shown in Fig. 3. The figure shows the degree distribution of all Ethereum addresses we extracted from the blockchain. The very few addresses with high degree of interaction are for example the mining pools and the exchanges.

Among the Ethereum addresses that interact with very few other addresses are the *one-time addresses* [12], which are only created for one transaction and are then never used again. The degree of an address over the selected time period can be found as an attribute of the address in EthExplorer, as we will discuss in the next section.

4   EthExplorer

This section introduces the software system in the first subsection and in the second subsection we explain how to use EthExplorer and what information it displays.

4.1   System Design

EthExplorer is a web-based graphical application to explore transaction data and
the flow of Ether and tokens based on Ether along the Ethereum blockchain. It
can be accessed at https://ethereum.imp.fu-berlin.de/.

The software architecture of the tool is quite simple. A web application serves
as interface with the user. The web application accesses the database through a
webserver which runs a Java application program. In essence, EthExplorer selects
information from the EthExplorer database and displays it using a graph layout.
As graph layouts we have included two algorithms from the Gephi Toolkit, the
Fruchterman–Reingold and the Yifan–Hu layout [13,14].

The graph layout algorithms are implemented in the Java library of the Gephi
Toolkit[1] [15], which is included in the Web application.

The most important component of EthExplorer is its database including the
annotations of the raw data. The database has been generated running a full
Parity mining node, which has downloaded the complete Ethereum blockchain
from other nodes in the P2P network. In this full node all blocks have been
scanned for transaction information. As can be seen in the database schema in
Fig. 4 the extracted components are the blocks, which can contain various transactions of three types: simple transactions, token transfers, and smart contract
transactions. Those transactions carry different attributes. The value of Ether in
Dollars at the time when a block is mined is saved in the block table (rate). In the
graphs the exchange rate between Ether and Dollar is used to determine the Dollar value of transactions. An important difference between our database entries
and the data on Ethereum nodes is that approximately 40% of the addresses

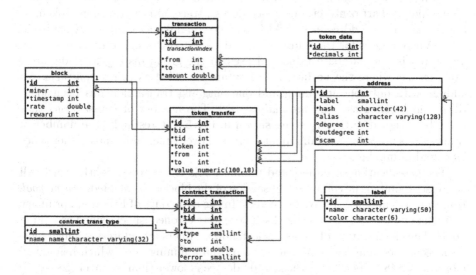

Fig. 4. Design of the Ethereum database

---

[1] https://gephi.org/toolkit/.

were labelled using information from other sources, such as Etherscan [8]. The size of the largest tables in the database is given in Table 1. We specify the number of rows in the table as well as the file size for the data and for the index information, which is larger than the pure data. The total file size of the tables is roughly 195 GB, which is significantly less than the 2000 GB needed by the full node to store the blockchain in archive mode and more than the 138 GB needed without archive mode. The runtime of a database query strongly depends on the type of query. A simple query which finds all transactions for one address needs 80–250 ms, while a complex query including ERC20/721 tokens and smart contract internal transactions can take minutes if the whole blockchain is traversed.

Table 1. Sizes of the database tables

	Block	Transaction	Address
Rows	6.047.918	279.807.270	67.953.422
Data	666 MB	14 GB	8563 MB
Index	789 MB	27 GB	28 GB
	contract_transaction	**token_transfer**	
Rows	511.408.420	102.261.273	
Data	33 GB	6094 MB	
Index	63 GB	11 GB	

Addresses were colour encoded if they belong to one of the following types: mining pool, smart contract, one time address, trace, miner, exchange, token, or the genesis block. Addresses for which no type could be determined are shown in grey. Where aliases for the addresses could be found through etherscan.io they are stored in the database table and used in the graph display as node label. Transactions where the to field is empty are classified as *smart contracts*. An address with only one incoming and one outgoing transaction has been identified and labelled as *one-time address* [12]. The number of transactions that could be assigned to the types are shown in Fig. 5. Interestingly, the number of transactions involving a mining pool decreased over time, while most other types increased in number.

The classification of miners and mining pools is not so straightforward. All addresses found in the coinbase transaction of the blocks are labelled *mining pool*. However, this may not always be correct. In the early days of Ethereum ordinary individual miners would mine blocks. Those were erroneously classified as mining pools. The classification of miners is as follows. Miners are addresses which have been recipients of at least one transaction from a mining pool, which has mined more than 2 489 blocks and whose node degree is lower than the average degree of all recipients of transactions from the pool. This definition uses two heuristics and will therefore not always be accurate. The first is to classify a miner as

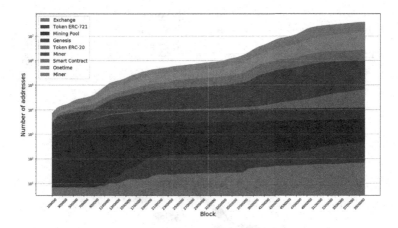

Fig. 5. Labelled addresses

the recipient of a mining pool which has mined more than 2 489 blocks. When discarding the first 1 000 000 blocks, the average number of blocks found by a mining pool, as classified above, is 2 489. Hence the selection of this threshold. The study of the node degree of recipients of nine popular mining pools, as shown in Fig. 12 confirms that most of those recipients have a low node degree. Hence we generalised this observation and assume that recipients of mining pools (as classified above) with a low node degree will be miners. We are aware that we may misclassify miners of the early days of Ethereum as mining pools, while they should rather be labelled as miners and we are possibly misclassifying some members of mining pools.

Exchanges were identified using their aliases on etherscan.io. To find the exchanges on etherscan.io we selected the top 500 addresses and their aliases according to node degree and verified the names manually. The ERC-721 tokens were identified using the token tracker website. The dollar value of a transaction is computed on the fly using the Ether value and exchange rate stored in the database.

4.2 How to Use EthExplorer

The web interface of EthExplorer is shown in Fig. 6. The web interface is still under development as its usability needs improvement. Upon accessing the web interface the display screen is empty. In the top panel the user can enter one or more Ethereum addresses or block number(s). A click on show/download will display the graph of the address or block. When entering a block number the transactions in the block will be shown as pairs of nodes (addresses) connected by an edge for the transaction. The thickness of the arc indicates the amount of the transaction in dollars at the time the transaction was issued. This is shown with the mouse over the arc. For clarity, nodes are not labelled with the corresponding Ethereum address by default. Instead, a node is labelled with the total number

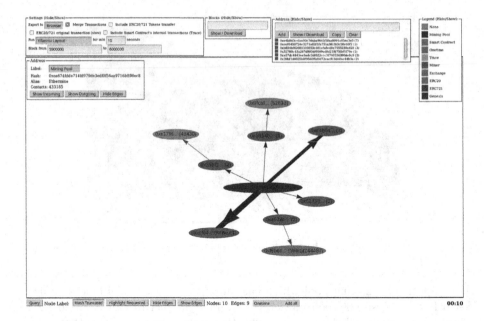

Fig. 6. Web interface of Ethexplorer

of neighbours it has, even if those are outside the graph display. Synonymous names are shown where they exist. The left most field in the bottom panel allows to select the hash identifier or an abbreviated hash as node label instead of the node degree.

Fig. 7. Settings block

The *Settings* block on the top left in the web interface, as shown in Fig. 7 allows to set parameters which can greatly improve the visual impression if the set of results is extremely large.

The settings allow to aggregate several transactions (edges) into one, which is then scaled by the total amount of the transactions. In addition, token transfers and activities of smart contracts as well as interaction with smart contracts can be included or excluded by selecting or un-selecting the respective buttons.

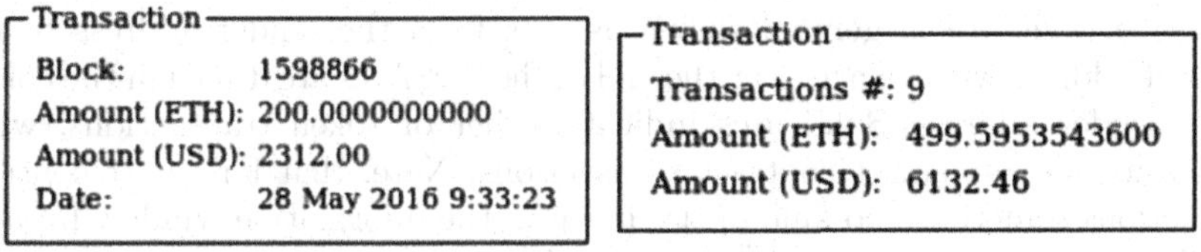

```
┌ Transaction ─────────────────────┐   ┌ Transaction ──────────────────────────┐
│  Block:            1598866        │   │  Transactions #: 9                     │
│  Amount (ETH): 200.0000000000     │   │  Amount (ETH):   499.5953543600        │
│  Amount (USD): 2312.00            │   │  Amount (USD):   6132.46               │
│  Date:             28 May 2016 9:33:23 │ │                                     │
└───────────────────────────────────┘   └────────────────────────────────────────┘
```

Fig. 8. Detailed information (left) and simple information (right) on a transaction

Clicking on an edge (a set of transactions) will open an information window as shown in Fig. 8 on the right, where the dollar value of the transactions has been determined as the sum of the individual transaction values at the time of their issuing.

The graph can either be shown in the browser, or it can be exported to Gephi [15] for further usage. The graph layout can use one of the two algorithms, Yifan–Hu layout [14] or Fruchterman–Reingold [13] algorithm. Yifan–Hu pushes nodes with low link count (degree) more strongly to the periphery, while Fruchterman–Reingold keeps a symmetrical sphere shape, offering a more harmonious picture. We mostly use the Fruchterman–Reingold layout, but depending on the subject of investigation the Yifan–Hu algorithm might more clearly show the centre of activities. We find that limiting the layout time to 10 s mostly provides a good result within reasonable time. Finally, the considered range of blocks can be limited to a subset of all available blocks in the database, which greatly speeds up the result, especially when a large graph is created.

More nodes can be shown by either entering addresses manually in the address field in the top panel, or by selecting an address type in the bottom panel on the right and adding all addresses of that type. This will add all shown nodes of that type to the list of explored addresses. Upon the next issue of show/download the direct neighbours of all those addresses will be added to the shown graph.

When unselecting the merge transactions button in the Settings section the click on an edge between two nodes will open an information window as shown in Fig. 8 on the left, where information on a transaction in block 1 598 866 is shown. The total amount of Ether transferred in this transaction is shown as well as its value and the time when the block holding this transaction was generated.

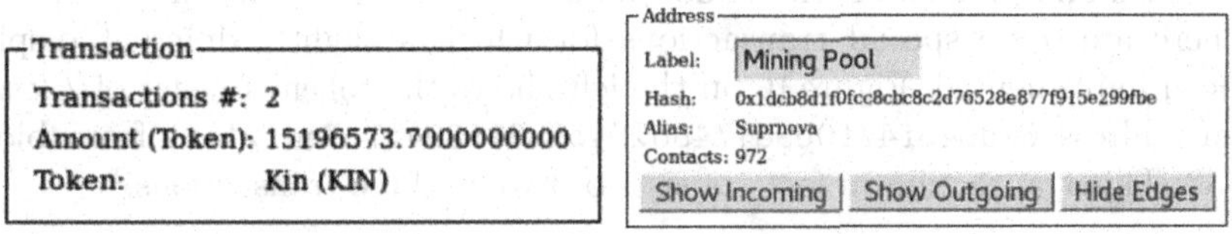

```
┌ Transaction ──────────────────────────┐   ┌ Address ────────────────────────────────────┐
│  Transactions #: 2                     │   │  Label:   │ Mining Pool │                      │
│  Amount (Token): 15196573.7000000000   │   │  Hash:    0x1dcb8d1f0fcc8cbc8c2d76528e877f915e299fbe │
│  Token:          Kin (KIN)             │   │  Alias:   Suprnova                           │
│                                         │   │  Contacts: 972                               │
│                                         │   │  [ Show Incoming ] [ Show Outgoing ] [ Hide Edges ] │
└─────────────────────────────────────────┘   └──────────────────────────────────────────────┘
```

Fig. 9. Information for a token transaction (left) and an address (right)

Transactions are in general shown as arcs from the sender address to the recipient address with an arrow at the end of the arc. Arcs are drawn in the colour of the sender address. Solid arcs indicate Ether or token transactions, while dashed arcs denote smart contract transactions. Note, that for smart contract transactions sometimes no amount is shown in the information window because not all smart contract transactions transfer Ether. Some of them transfer tokens, which is then shown in the information window. In the latter case the mouse over the edge shows $0.0, while for Ether transfer it shows the corresponding dollar amount.

Moving the mouse over a node will show the dollar amount of an edge originating at that node. The balance of an address can not yet be shown by EthExplorer, this feature is among future work.

Selecting an arc will open up an information window as shown in Fig. 8 on the right, which provides information on the number of transactions joined in the arc and their total accumulated value, each at the time of issuing the transaction. If the selected arc belongs to a token transaction the information window, as shown in Fig. 9 on the left will display information on the type and amount of token transferred in the accumulated transactions. In the shown example two transactions were joined in one arc carrying a total of 15 196 573.7 Kin, a token connected to the Kik messenger app (https://www.kin.org/).

Fig. 10. Graphical presentation of an ERC20/721 transaction (left) and internal transaction of a smart contract (right) (Color figure online)

A mouse click on a blue node will open the address information window as shown in Fig. 9 on the right. In general, information windows for nodes come with buttons to remove either in- or outgoing arcs, or both. In large graphs this can be very helpful. Removing arcs from the display of the overall graph can be done using the respective buttons in the bottom panel of the web interface. A double click on a node adds the node's address to the list of shown addresses.

There are some special transactions for which a slightly different display has been implemented. Figure 10 on the left shows the token transfer of *Elcoins* (token address 0xa04bf47f0e9d1745d254b9b89f304c7d7ad121aa) from block 917 132. This transaction in fact consists of two or three transactions.

Assume a token is transferred from address A to address B, then selecting the parameter *ERC20/721 original transaction (slow)* in the settings window on the top left will show an additional transaction from address C to D which has triggered the token transfer. Sometimes a token transfer is triggered by two transactions, as in Fig. 10 on the left[2]. The same transaction appears in etherscan.io as shown in Fig. 11.

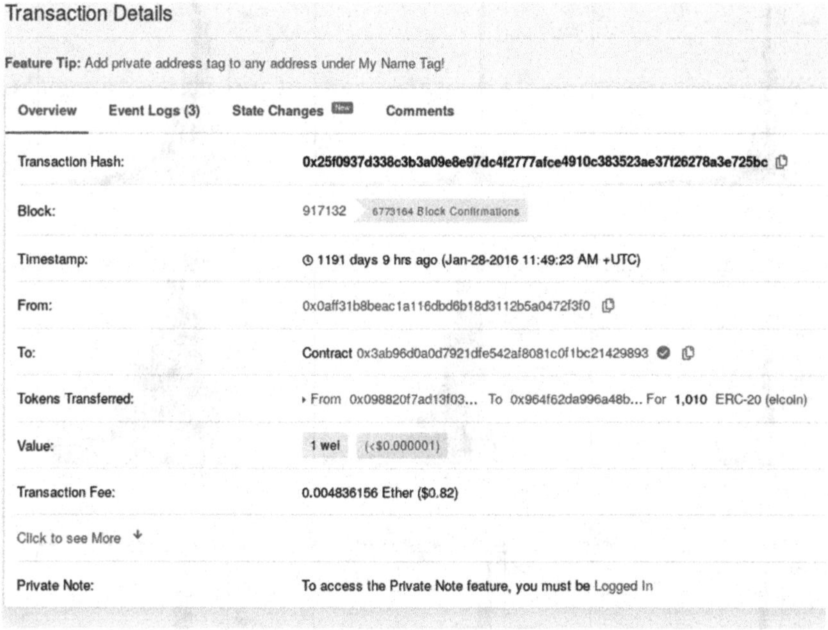

Transaction Details

Feature Tip: Add private address tag to any address under My Name Tag!

Overview	Event Logs (3)	State Changes	Comments

Transaction Hash:	0x25f0937d338c3b3a09e8e97dc4f2777afce4910c383523ae37f26278a3e725bc
Block:	917132 6773164 Block Confirmations
Timestamp:	① 1191 days 9 hrs ago (Jan-28-2016 11:49:23 AM +UTC)
From:	0x0aff31b8beac1a116dbd6b18d3112b5a0472f3f0
To:	Contract 0x3ab96d0a0d7921dfe542af8081c0f1bc21429893
Tokens Transferred:	▸ From 0x098820f7ad13f03... To 0x964f62da996a48b... For 1,010 ERC-20 (elcoin)
Value:	1 wei (<$0.000001)
Transaction Fee:	0.004836156 Ether ($0.82)
Click to see More ↓	
Private Note:	To access the Private Note feature, you must be Logged In

Fig. 11. The ERC20/721 transaction in etherscan.io

In such cases the arrows of the secondary transaction are not drawn at the destination, but in the middle of the arc. Including such transactions in the search slows down the database request considerably, which is therefore indicated on the respective button.

Another special type of transaction are the smart contract internal transactions as shown in Fig. 10 on the right. A mining pool issues a transaction to a smart contract which is then executed with an output to an unlabelled (grey) address of unknown type.

---

[2] See https://etherscan.io/tx/0x25f0937d338c3b3a09e8e97dc4f2777afce4910c383523ae 37f26278a3e725bc for the transaction(s) shown in the figure.

5   Observations on the Ethereum Blockchain

In face of the large number of Ethereum blocks, transactions and addresses the real challenge is to be able to filter and see the relevant pieces of information within the flood of data.

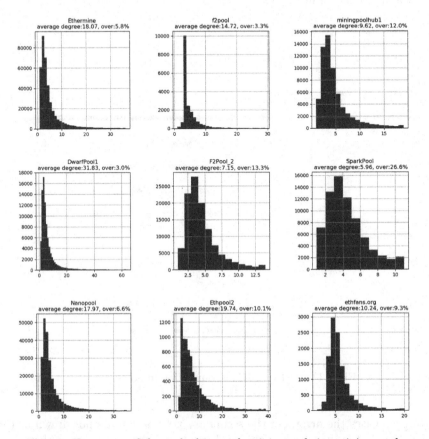

Fig. 12. Histogram of the node degree of recipients of nine mining pools

Mining pools provide an easy interface to miners, who very often use their address for mining, receiving their reward and transferring it to an exchange or another destination for payout. This assumption is supported by the analysis of the node degree of the recipients of the nine most popular mining pools as shown in Fig. 12.

Most participants in mining pools have only very few contacts. For mining and payout two contacts are needed and most members of the mining pools indeed interact with only few other addresses, less than 10 addresses in total. This is very different for the mining pools themselves, for exchanges or other addresses that provide services.

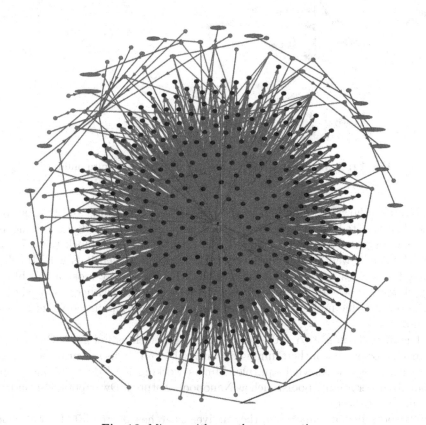

Fig. 13. Miners with no other transactions

For the beauty of presentation Fig. 13 shows a set of 390 miners, labelled as mining pools, having mined a total of 4223 blocks, but who have not made any other transactions. Such sets can be determined with a direct query to the database using the `query` button to open a field for database requests. Interestingly, those miners have instead bought 35 different tokens (mostly `An Etheal Promo`, `INS Promo`, `VIU`, `XENON`, `Datacoin`, `OmiseGO` and `BitClave`).

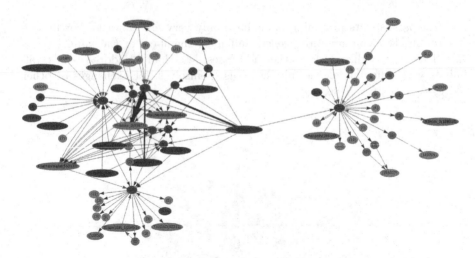

Fig. 14. Ten miners of Ethermine (in violet) (Color figure online)

Figure 14 shows ten members of the mining pool Ethermine and their activities. The selected addresses have performed transactions with Ethermine in a volume between 0.1 and 5 Ether. Until autumn 2017 the block reward was 5 ETH, then it was 3 ETH until February 2019, since then it is 2 ETH. The selected addresses made transactions with Ethermine of at most a full block reward, so they are all considered small miners. In Fig. 14 Ethermine is shown as the central blue node with ten outgoing arcs. The arcs point to violet nodes, the miners.

The miners connect to addresses of exchanges, such as Poloniex, Bitfinex or Changelly for payout of their mining reward. Some miners use several exchanges. The exchanges are shown in purple. Miners working for Ethermine also contribute to other mining pools, such as Nanopool, Antpool, Dwarfpool, also shown in blue, like Ethermine.

The bottom left miner is rather active, as it has degree 19. It trades on *ShapeShift* and invests in *The DAO,* a token that aims at creating a decentralised organisation. The miner on the right has 22 contacts, among them three exchange sites, another miner (potentially another address used by the same person), a number of *one time addresses*, and several unlabelled addresses for which no information exists. We are wondering whether one-time addresses are implemented by some exchanges as gateway addresses into the exchange, used for single payouts.

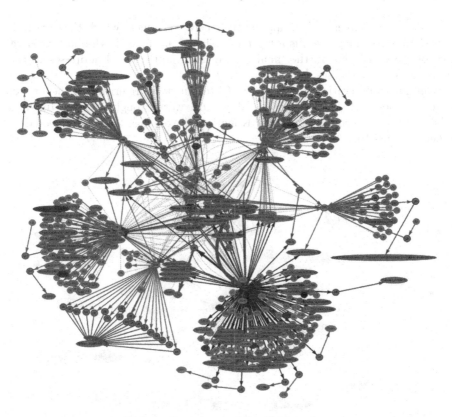

Fig. 15. Recipients of the Golem token

Figure 15 shows another study looking at the recipients of the ERC-20 Golem token (GNT). Golem is a system where participants can offer their computing resources for usage by others in exchange for Golem tokens. The graph has 556 nodes and 1 039 edges. The Ethereum address of Golem can be found on etherscan.io and the following database query delivers the shown nodes, which are then enriched by adding all one-time addresses using the button in the bottom panel of EthExplorer.

```
SELECT t.to
FROM token_transfer t
JOIN address a ON t.token = a.id
WHERE a.hash = lower(
0xa74476443119A942dE498590Fe1f2454d7D4aC0d)
LIMIT 10
```

Golem is shown as one of the three turquoise nodes left of the centre at around 10 o'clock. It is the token node with eight outgoing edges. Interestingly, the recipients of Golem tokens use many other tokens as well and the connected one-time addresses link to addresses with very high degree, such as exchanges.

Users of the Golem token apparently receive payments which they transfer to exchanges. Among them is a big miner shown in purple in the bottom right part of the graph. For further analysis we could add selected addresses to the list of explored addresses, but we skip that here.

Our final observation concerns the Gatecoin hack, referring to the loss of 185 000 Ether and 250 Bitcoins in May 2016 worth roughly US$2.14m in a cyber attack. Gatecoin was a Hong Kong based exchange loosely connected to Ethereum-based DAOs.

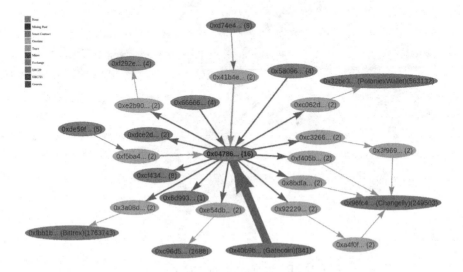

Fig. 16. Hack of Gatecoin

Figure 16 shows a large transfer from a Gatecoin address to an unlabelled one from where several smaller amounts were transferred through one-time addresses and unlabelled addresses to other exchanges, i.e. Poloniex, Bittrex and Changelly.

6   Conclusion

We have presented EthExplorer, a tool for the analysis of activities on the Ethereum blockchain and the flow of Ether between addresses. The tool allows to investigate the transactions made by given addresses and to successively expand and reduce the set of explored addresses. EthExplorer uses a database, which we have generated while running a Parity full node and enriched with additional information, such as the type of an address and the dollar value of a transaction at its time of issue. The graph layout is taken from Gephi, where we use two standard algorithms, the Fruchterman–Reingold layout and the Yifan–Hu algorithm. The selected addresses and their transactions can either be displayed in a web interface, or can be exported to Gephi for further processing.

EthExplorer is a powerful tool, but still many issues remain open. Currently it is not possible to determine the balance of an address. Only amounts of transaction transfers are stored, converted and accumulated. The graph display can also be further improved by adding a selection slider based on calendar time, rather than on block number. Finally, when changing the list of explored addresses the graph layout is triggered, removing a possibly better manual layout. It should be possible to maintain a layout and simply add or remove nodes and edges. Last, we are working on an online update, to keep the database up to date at all times.

Acknowledgments. We would like to thank the anonymous reviewers for their insightful comments on the paper, especially for the suggestion to make our database directly available to the public. We will work on this.

References

1. Bulterin, V.: A next-generation smart contract and decentralized application platform (2013)
2. Wood, G.: Ethereum: a secure decentralised generalised transaction ledger (2018)
3. Nakamoto, S.: Bitcoin: a peer-to-peer electronic cash system (2008)
4. O'Kane, E.: Detecting patterns in the Ethereum transactional data using unsupervised learning. Master's thesis, University of Dublin, Trinity College, Dublin, Ireland, August 2018
5. Chan, W., Olmsted, A.: Ethereum transaction graph analysis. In: 2017 12th International Conference for Internet Technology and Secured Transactions (ICITST), pp. 498–500 (2017)
6. Chen, T., et al.: Understanding ethereum via graph analysis. In: IEEE INFOCOM 2018 - IEEE Conference on Computer Communications, pp. 1484–1492, April 2018
7. McGinn, D., Birch, D., Akroyd, D., Molina-Solana, M., Guo, Y., Knottenbelt, W.J.: Visualizing dynamic bitcoin transaction patterns. Big Data **4**(2), 109–119 (2016)
8. How Can I Add My Name Next To Address On Etherscan? (2017). https://www.reddit.com/r/ethereum/comments/4d612u/how_can_i_add_my_name_next_to_address_on_etherscan/d1p8ns9/. Accessed 10 Jan 2019
9. Zamyatin, A., Wolter, K., Werner, S., Harrison, P.G., Mulligan, C.E.A., Knottenbelt, W.J.: Swimming with fishes and sharks: beneath the surface of queue-based Ethereum mining pools. In: 2017 IEEE 25th International Symposium on Modeling, Analysis, and Simulation of Computer and Telecommunication Systems (MASCOTS), pp. 99–109, September 2017
10. Vogelsteller, F., Buterin, V.: ERC-20 token standard, 2015 (2018). https://github.com/ethereum/EIPs/blob/master/EIPS/eip-20-token-standard.md
11. Entriken, W., Shirley, D., Evans, J., Sachs, N.: Non-fungible token standard, document ERC-721, September 2018
12. Huge ethereum mixer. Accessed 19 July 2018
13. Fruchterman, T., Reingold, E.: Graph drawing by force-directed placement (1991)
14. Hu, Y.F.: Efficient and high quality force-directed graph drawing. Math. J. **10**, 37–71 (2005)
15. Bastian, M., Heymann, S., Jacomy, M.: Gephi: An open source software for exploring and manipulating networks (2009)

A Queueing Model that Works Only on the Biggest Jobs

Andrea Marin[⊠] and Sabina Rossi

Università Ca' Foscari Venezia, via Torino, 155, 30172 Venice, Italy
{marin,sabina.rossi}@unive.it

Abstract. We consider a queueing system with capacity 1 and subject to a Poisson arrival process. Jobs consists of a random number of tasks and at each arrival, the system will continue to work on the current job if the number of its tasks is higher or equal than the number of tasks of the job just arrived, otherwise the job in the queue leaves the system and the one just arrived begins its service. The service time of each task is independent and exponentially distributed with the same parameter.

We give an explicit solution for the stationary distribution of the queue by resorting to time-reversed analysis and we observe that this approach gives a much more elegant and constructive way to obtain the result than the traditional approach based on the verification of the system of global balance equations. For geometric distribution of the number of tasks, we use the q-algebra to make the results numerically tractable. The queueing system finds applications in contexts in which the size of jobs is known or partially known and schedulers or dispatchers can take decisions based on this information to improve the overall performance (e.g., reducing the mean response time).

Keywords: Queueing systems · Reversed-time analysis · q-series

1 Introduction

In the last decades, queueing models that can take advantage of the exact or approximate knowledge of the job sizes have been widely investigated (see, e.g., [2,5,10,13]). Applications scenarios in which some form of knowledge of the job sizes is possible includes the scheduling of TCP flows from a web servers (see, e.g., [15]) or other scenarios in which the precise size of a job is unknown, but the number of tasks which must be performed to complete its service is known [10,13]. Although the main application of size-based scheduling is the implementation of disciplines that mimic or implement the *Shortest Remaining Processing Time (SRPT)* thanks to its optimality in the average response time [14], in this paper, we use similar ideas with a different purpose, i.e., we aim at maximising the system utilisation.

That system that we presented here is similar to those studied in [3,12], where queueing networks consisting of finite capacity stations follow the *skipping policy*:

© Springer Nature Switzerland AG 2020
M. Gribaudo et al. (Eds.): EPEW 2019, LNCS 12039, pp. 118–132, 2020.
https://doi.org/10.1007/978-3-030-44411-2_8

at the arrival of a job at a saturated queue, this is directly (probabilistically) routed to the next station. In our case, we complicate the routing decision by comparing the size of the job in service and that of the just arrived job. Given the difficulties in the exact analysis, we limit our study to a single queueing system since, in contrast with the models studied in [3,12], the composition of several components of such a type is not in product-form.

Let us consider a queueing system with capacity 1. Job arrives according to a time-homogeneous Poisson process and consist of a random number of tasks; we will use the term *job size* to denote the number of tasks it consists of, while we use *job length* to refer to the sum of the sizes of all the tasks that contribute to its service time. The size of each task is independent an exponentially distributed. At each arrival, the queue will begin the service on the job with larger remaining size, while the other leaves the system (e.g., may migrate to a slower queue).

The contributions of this paper can be summarised as follows:

- We give an explicit expression of the invariant measure of the queue. For the case of geometric job sizes, we also give the expression of the normalising constant and prove that the system is unconditionally stable. From the theoretical point of view, the case of geometric job sizes is particularly interesting since it involves some beautiful results inherited from the q-series analysis.
- We compare the utilisation of the system that works only on the largest jobs with that of a system that is unaware of the job sizes such as those considered in [3,12].
- We also emphasise the benefits of an analysis based on the time-reversed chains. Indeed, we give the proof of the stationary distribution for the model with geometric job sizes by using the global balance equation approach, while the proof for the case of general independent distributions is based on the analysis of the time-reversed chain. Clearly, the former may be derived by the latter as a special case (and we show, as sanity check, that this is possible), but we believe that the two versions improve the readability of the paper by allowing researcher not familiar with time-reversed analysis to access to the results and helps in appreciating the constructive approach of the latter proof method.
- Finally, we show the connections of this queueing system, specifically of its stationary idle probability, with the generating function of number sequence A008289 [1] which plays an important role in combinatorics.

The paper is structured as follows. Section 2 reviews the literature related to this contribution. In Sect. 3 we introduce and solve the queueing model with geometrically distributed job sizes while the general case is considered in Sect. 4. Section 5 illustrates the connections between these results and number theory. Finally, Sect. 6 concludes the paper.

2   Related Work

The idea of modelling the partial knowledge of the job length by means of the knowledge of the tasks it consists of can be found in [10,13]. In [13] the authors consider a queueing system with infinite capacity that modulates its service speed according to the size of the batch it is serving. In [10], the authors study the SRPT scheduling discipline with speed scaling by approximating the exact knowledge of the job length with the exact knowledge of the number of tasks that form the job. In [3,12], the authors consider the *skipping policy* for handling finite capacity queueing systems and prove that a network of queues with such a discipline is in product-form. In our case, the *skipping policy* is more complicated because it involves the comparison of the job sizes, and so is the analysis. As unfortunate consequence, we loose the property of separability in the stationary distributions of networks of this type of queues.

The results on time-reversed analysis that we use in our proofs can be found in [6–9].

3   The Queueing Model

We consider a queueing model that can store at most one job. Each job consists of a finite but potentially unbounded number of tasks, thus the capacity of the queue for the tasks is infinite. The jobs arrive from the outside according to an independent and homogeneous Poisson process. We assume that each job consists in a random number B of tasks. We call B the *size of the job*, in contrast with its *length*, which is instead the sum of the sizes of the tasks. In this section, we consider the size of the jobs to be geometrically distributed:

$$Pr\{B = k\} = (1 - \beta)\beta^{k-1} = p_k \,,$$

where $0 < \beta < 1$, while in Sect. 4 we generalise the result for arbitrary distributions of the job sizes. The only way that the system has to estimate the remaining work on a customer is given by the number of tasks that still have to be served. Each task has an independent exponentially distributed size whose parameter is $\mu \in \mathbb{R}^+$. Thus, the length of a job J is exponentially distributed with average:

$$E[J] = \frac{1}{(1 - \beta)\mu},$$

Since the sum of a geometric number of independent and identically distributed exponential random variables is an exponential random variable. The server has constant unitary speed.

The system aims at working on the job that requires more refinements, i.e., the job with the largest amount of remaining tasks. Let us assume that at time t the system is working on a job consisting of $n(t)$ remaining tasks, and that a job of size B arrives. Then, immediately after the arrival epoch, t^+, the state is:

$$n(t^+) = \max(B, n(t)).$$

Thus, a job may leave the system for two reasons: either because the server has completed its work or because a job with a larger amount of tasks arrived from the outside. Since we assume that the arrival process of jobs is a homogeneous

Poisson process with intensity λ, then $n(t)$ is a continuous time Markov chain (CTMC) whose state space is $\mathbb{N}$.

The transition rates for the CTMC are defined as follows:

$$q(n_1, n_2) = \begin{cases} \mu & \text{if } n_2 = n_1 - 1 \text{ and } n_2 \geq 0, \\ \lambda p_k & \text{if } n_2 = k \text{ and } k > n_1. \end{cases} \tag{1}$$

3.1 Stationary Analysis of the Model

In this part, we propose the derivation of the expression of the stationary probability distribution. We will see that its explicit form relies on some q-series analysis. First, let us introduce the system of global balance equations (GBE) for the queue:

$$\begin{cases} \pi(i) \left(\mu + \lambda \sum_{j=i+1}^{\infty} p_j \right) = \sum_{j=0}^{i-1} \pi(j) \lambda p_i + \pi(i+1)\mu, & i > 0 \\ \pi(0)\lambda = \pi(1)\mu \end{cases} \tag{2}$$

which can be rewritten as:

$$\begin{cases} \pi(i) \left(\mu + \lambda \beta^i \right) = \lambda(1-\beta)\beta^{i-1} \sum_{j=0}^{i-1} \pi(j) + \pi(i+1)\mu, & i > 0 \\ \pi(0)\lambda = \pi(1)\mu \end{cases}$$

Let us introduce the following Lemma that gives the invariant measure for the queueing system, while Theorem 1 will discuss the stability of the queue and gives its stationary distribution.

Lemma 1. *The invariant measure for the queueing system described by the infinitesimal generator (1) is*

$$g(n) = \begin{cases} \kappa \rho \beta^{n-1} (-\rho; \beta)_{n-1} & \text{if } n > 0 \\ \kappa & \text{if } n = 0 \end{cases}, \tag{3}$$

where $\kappa \in \mathbb{R}^+$, $\rho = \lambda/\mu$, and $(a; q)_n$ is the q-Pochhammer's symbol:

$$(a; q)_n = \prod_{k=0}^{n-1} (1 - aq^k).$$

We present two proofs of Lemma 1. The first is based on verifying that Eq. (3) satisfies the system of global balance Eq. (2). This proof is not constructive and, differently from what usually happens with Markovian queues of the type M/M/k, neither the formulation of the 'educated guess' on the expression of the invariant measure nor the algebraic steps to prove its correctness are simple or intuitive. In contrast, in Sect. 4 we consider batches of arbitrary size distribution and we propose a constructive method to quickly derive the expression of the stationary distribution based on time-reversed analysis. The model with geometrically distributed job sized is a special case.

Proof. By replacing Expression (3) in the first equation of System (2) we have:

$$\kappa\rho\beta^{i-1}(-\rho,\beta)_{i-1}(\mu+\lambda\beta^i)$$

$$= \lambda(1-\beta)\beta^{i-1}\kappa\rho\left(\sum_{j=1}^{i-1}\beta^{j-1}(-\rho,\beta)_{j-1}\right) + \lambda(1-\beta)\beta^{i-1}\kappa + \kappa\rho\beta^i(-\rho,\beta)_i\mu.$$

Let us divide both hand sides of the equation by $\kappa\rho(-\rho,\beta)_{i-1}$ which is strictly positive since the empty product gives 1. Then, we have:

$$\mu+\lambda\beta^i = \frac{\lambda(1-\beta)}{(-\rho,\beta)_{i-1}}\sum_{j=1}^{i-1}(-\rho,\beta)_{j-1}\beta^{j-1} + \frac{1-\beta}{(-\rho,\beta)_{i-1}}\mu + \beta(1+\beta^{i-1}\rho)\mu,$$

which can be reduced to:

$$\mu = \frac{\lambda(1-\beta)}{(-\rho,\beta)_{i-1}}\sum_{j=1}^{i-1}(-\rho,\beta)_{j-1}\beta^{j-1} + \frac{(1-\beta)}{(-\rho,\beta)_{i-1}}\mu + \beta\mu.$$

Therefore, we need to prove that:

$$\mu = \frac{\lambda}{(-\rho,\beta)_{i-1}}\sum_{j=1}^{i-1}(-\rho,\beta)_{j-1}\beta^{j-1} + \frac{1}{(-\rho,\beta)_{i-1}}\mu. \tag{4}$$

We proceed by induction on i. The case $i = 1$ trivially gives an identity since the first sum is 0 and $(-\rho,\beta)_0 = 1$. Let us consider the case $i+1$, with $i \geq 1$, then we have:

$$\frac{1}{(-\rho,\beta)_i}\left(\lambda\sum_{j=1}^{i}(-\rho,\beta)_{j-1}\beta^{j-1} + \mu\right)$$

$$= \frac{1}{(-\rho,\beta)_{i-1}}\frac{(-\rho,\beta)_{i-1}}{(-\rho,\beta)_i}\left(\lambda\sum_{j=1}^{i-1}(-\rho,\beta)_{j-1}\beta^{j-1} + \lambda(-\rho,\beta)_{i-1}\beta^{i-1}\mu\right)$$

$$= \frac{1}{(-\rho,\beta)_{i-1}}\frac{\beta}{\beta+\rho\beta^i}\left(\lambda\sum_{j=1}^{i-1}(-\rho,\beta)_{j-1}\beta^{j-1} + \mu\right) + \frac{\lambda\beta^i}{\beta+\rho\beta^i}.$$

By inductive hypothesis, we can rewrite the expression as:

$$\mu\frac{\beta}{\beta+\rho\beta^i} + \frac{\lambda\beta^i}{\beta+\rho\beta^i},$$

which easily simplifies to μ as required by Eq. (4). □

Theorem 1. *The queue described by the infinitesimal generator (1) is unconditionally stable. The steady-state distribution has the following expression:*

$$\pi(n) = \pi(0)\rho\beta^{n-1}(-\rho;\beta)_{n-1}, n > 0 \tag{5}$$

where $0 < \pi(0) < 1$ is the probability of the empty queue:

$$\pi(0) = \frac{1}{(-\rho; \beta)_\infty}.$$

Proof. Thanks to Lemma 1 we just need to focus on the computation of $\pi(0)$. If we can prove that $0 < \pi(0) < 1$ for $0 < \beta < 1$, then the CTMC is ergodic since its transition matrix is irreducible and at least state 0 is positive recurrent. By definition, we have $\pi(0) = (1 + G)^{-1}$, where:

$$G = \sum_{n=1}^{\infty} \rho\beta^{n-1} (-\rho; \beta)_{n-1}. \tag{6}$$

Lemma 2 allows us to conclude the proof about the stability of the chain and the expression of $\pi(0)$. Indeed, since in stability $G = -1 + (-\rho; \beta)_\infty$, we have that $\pi(0) = 1/(-\rho; \beta)_\infty$. $(-\rho, \beta)_\infty > 1$ is convergent since we can rewrite the product as:

$$\exp\left(\sum_{i=0}^{\infty} \log(1 + \rho\beta^i)\right)$$

and, by D'Alambert convergence criterion:

$$\lim_{i \to \infty} \frac{\log(1 + \rho\beta^{i+1})}{\log(1 + \rho\beta^i)} = \beta < 1.$$

Moreover $(-\rho; \beta)_\infty > 1$ since it is the product of infinite terms strictly greater than 1. Thus $0 < \pi(0) < 1$, as required. $\square$

Lemma 2. *If $0 < \beta < 1$, the following relation holds:*

$$\sum_{n=1}^{\infty} \rho\beta^{n-1}(\rho; \beta)_{n-1} = -1 + (-\rho, \beta)_\infty.$$

Proof. Let us introduce the following equality assuming the convergence of the series and products present:

$$-1 + \prod_{i=1}^{\infty}(1 + a_i) = \sum_{i=1}^{\infty} a_i \prod_{j=1}^{i-1}(1 + a_i), \tag{7}$$

which can be easily proved by rearranging the terms of the product:

$$-1 + (1 + a_1)(1 + a_2)(1 + a_3) \cdots,$$

as

$$a_1(1) + a_2(1 + a_1) + a_3(1 + a_1)(1 + a_2) + \ldots.$$

Now, let us consider this latter expression and observe that it corresponds to the definition of G according to Eq. (6) by setting $a_i = \rho\beta^{i-1}$. Moreover, it is easy to see that this series converges for $\rho > 0$ and $0 < \beta < 1$, as required. Therefore, we can write:

$$G = -1 + \prod_{i=1}^{\infty}(1 + \rho\beta^{i-1}),$$

where the second addend is the definition of $(-\rho; \beta)_\infty$. □

Example 1. Let us consider the system with jobs of size 1. In this case, the queue can contain 1 or 0 tasks. The transition rate from 1 to 0 is μ and that from 0 to 1 is λ. Thus, we have $\pi(0) = \mu/(\lambda + \mu)$ and $\pi(1) = \lambda/(\lambda + \mu)$. If we consider Theorem 1 and take the limit $\beta \to 0^+$, we obtain $\pi(0) = (1 + \rho)^{-1} = \mu/(\lambda + \mu)$ and $\pi(1) = \pi(0)\rho = \lambda/(\lambda + \mu)$, with all the other states with negligible probability, as expected.

Unfortunately, $\pi(0)$ does not have a closed form expression. However, we can approximate it by resorting to the Pochhammer's symbol approximation. In [11, Thm. 4] an accurate approximation is presented under the conditions $\rho > 0$ and $0 < \beta < 1$, i.e., we have:

$$\log\left((-\rho_1, \beta)_\infty\right) = \sum_{k=-1}^{p} \frac{(-1)^k B_{k+1} L_{1-k}(-\rho_1)}{(k+1)!}(-\log(\beta))^k + O((-\log(\beta))^{p+1}),$$

where:

- B_k is the k-th Bernoulli number whose exponential generating function is:

$$g_B(x) = \sum_{k=0}^{\infty} B_k \frac{x^k}{k!} = \frac{x}{e^x - 1},$$

- L_n is the n-th Polylogarithm function:

$$L_k(z) = \sum_{j=1}^{\infty} \frac{z^j}{j^k},$$

that for negative integer values of k can be expressed by rational functions as:

$$L_{-n}(z) = \left(z\frac{\partial}{\partial z}\right)^n \frac{z}{1-z}, n \geq 0,$$

while:

$$L_0(z) = \frac{z}{1-z}, \quad L_1(z) = -\log(1-z), \quad L_2(z) = -\int_0^z \frac{\log(1-t)}{t}dt.$$

As we can see, the approximation becomes quite accurate when $\beta \to 1^-$. For practical purposes, we can have very low values of p to obtain very good approximations. For instance for $p = 2$ we have:

$$\pi(0)^{-1} \simeq (\pi^*(0))^{-1} = e^{L_2(-\rho)/\log(\beta)}\sqrt{1+\rho}\,\beta^{-\frac{\rho}{12(1+\rho)}}. \tag{8}$$

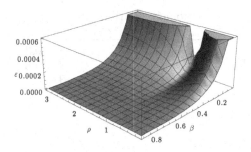

Fig. 1. Approximation of π_0: relative error as function of the parameters ρ and β.

In Fig. 1, we show the error between the approximation given by Eq. (8) and the approximation of the q-Pochhammer symbol obtained by the series expansion of Mathematica. This software approximates the q-Pochammer symbol at arbitrary precision, although the algorithm is not really applicable for symbolic analysis. We define ε as the relative error:

$$\varepsilon = \frac{|\pi^*(0) - \pi(0)|}{\pi(0)},$$

Under the assumption that $\pi(0)$ is evaluated at the highest machine precision. As we may observe, the approximation tends to become worse when $\beta \to 0^+$ and ρ is high. However, this is also the case when the system becomes 'less interesting' since the jobs are mainly formed by single tasks and hence in most of the cases have the same size and find the system either empty of with a job with their same size.

Example 2. Let us consider a system with $\lambda = 0.8$, $\mu = 1$. In Fig. 2 we show the stationary probability distribution for three values of β, while in Fig. 3 we show the impact of μ for fixed values of $\lambda = 0.8$ and $\beta = 0.4$.

As in many other queueing systems, the idle probability $\pi(0)$ plays a crucial role in the performance analyses. Figure 4 shows the stationary probability of finding an empty queue. As expected, the idle probability tends to 0 when the size of the jobs or the arrival rate increase.

Example 3 (Comparison with a queue with blocking). In this example, we compare the idle stationary probability of the queue studied in this section and the simpler one, $\pi_b(0)$, that accepts a job if it is empty or drops it if it contains one job. This queue simply is a M/M/1 queue with finite capacity (1) and rejection. The length of the jobs are exponentially distributed with expectation $((1 - \beta)\mu)^{-1}$. Then, we have:

$$\pi_b(0) = \frac{\mu(1 - \beta)}{\lambda + \mu(1 - \beta)}.$$

In Fig. 5, we show the ratio between $\pi_b(0)$ and $\pi(0)$. We observe that the queueing system that works only on the largest jobs gives a much better utilisation when β

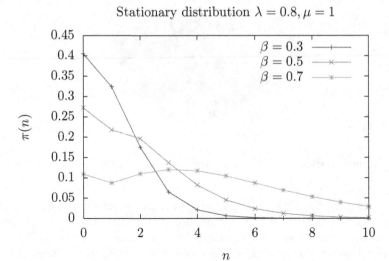

Fig. 2. Stationary distribution of the number of tasks in the queue for different values of β.

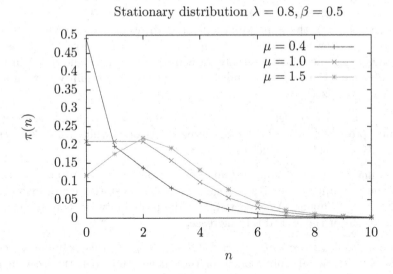

Fig. 3. Stationary distribution of the number of tasks in the queue for different values of μ.

is close to 1 (jobs consisting of many tasks) and high load factors. This is rather intuitive since we have already observed that, when the jobs consist of few tasks, the finite capacity queue and that studied here have the same behaviour (see Example 1). Moreover, when the load factor is low, the system tends to complete

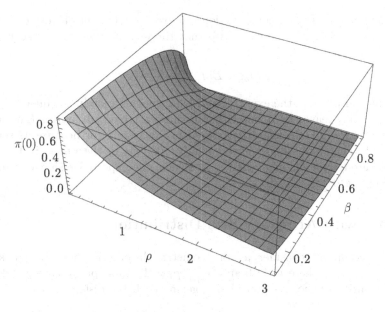

Fig. 4. Idle probability as function of μ and β.

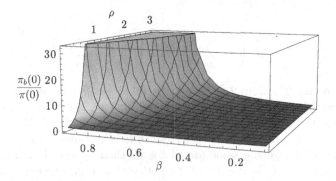

Fig. 5. Comparison of the idle probabilities of a M/M/1/1 queue with blocking and the queue working only on the largest jobs.

the service on the jobs and hence the replacement mechanism has a small impact on the stationary idle probability.

The work performed by the system in stability is $1 - \pi(0)$ jobs per unit of time (recall that we assume that the server has a constant speed of 1 job per unit of time), thus if the expected size of the jobs in input is $E[J]$, the expected amount of work arriving at the system in an long time interval Δt is:

$$E[J]\lambda\Delta t = \frac{\lambda\Delta t}{(1-\beta)\mu}.$$

In the same interval, the queue has done an expected amount of work expressed by $(1 - \pi(0))\Delta t$. Thus, we can write that the expected length of the jobs in output, $E[J_o]$, is:

$$E[J_o] = E[J] - \frac{1 - \pi(0)}{\lambda}. \tag{9}$$

We may also derive the expected residence time of a job in the queue. Recall that a job may leave the queue either because a larger job arrives (i.e., a job consisting of more tasks than that stored in the system) or because its service is finished. Therefore, the expected number of jobs in the queue is $1 - \pi(0)$ and the throughput of the system is λ. Hence, the expected residence time is: $E[R] = (1 - \pi(0))/\lambda$.

4   Jobs with Arbitrary Size Distribution

In this section, we consider a non-geometric distribution for the job sizes: $p_1, p_2, \ldots, p_n, \ldots$, where $p_b \geq 0$ and $\sum_{j=1}^{\infty} p_j = 1$. As a special case, we derive the distribution of the case considering geometric batch distribution.

Theorem 2. *In stability, the queueing model with arbitrary job size distribution has the following stationary distribution:*

$$\pi(n) = \pi(0)\frac{\lambda}{\mu^n}\left(1 - \sum_{j=1}^{n-1} p_j\right) \prod_{j=0}^{n-2}\left(\lambda\left(1 - \sum_{k=1}^{j} p_k\right) + \mu\right). \tag{10}$$

Before proving Theorem 2, it may be useful to recall some important results on time-reversed analysis. These can be found in [6–8]. Let $X(t)$ be a stationary CTMC, then $X(\tau - t)$ is also a stationary CTMC for all $\tau \in \mathbb{R}$. Henceforth, we will call $X(\tau - t)$ as $X^R(t)$ for brevity (since it is stationary). If the statistics of $X(t)$ and $X^R(t)$ are the same for all $t \in \mathbb{R}$, then we say that $X(t)$ is time-reversible. However, it is important to notice that even stationary processes that are not reversible admit a time-reversed process. This is the case for this queue that, in fact, does not have an underlying reversible CTMC. Generalised Kolmogorov's criteria provide necessary and sufficient conditions for deciding if $Y(t)$ is the reversed process of $X(t)$ based on the analysis of the transition rates.

K1: For every state of the chain, its residence times in the forward and reversed processes have the same distribution;

K2: For every cycle $i_1 \to i_2 \to \cdots \to i_n \to i_1$, with i_k a state of the CTMC and $n \geq 2$, the product of its rates in the forward chain is equal to the product of the rates in the reversed process.

Henceforth, we denote the transition rate from state i to state j by q_{ij}, and the rate of the inverse transition by q_{ji}^R. Why is it so important to know the rates of the reversed process? Indeed, it turns out that if we know the rates of the forward and reversed processes, then we can easily compute the invariant

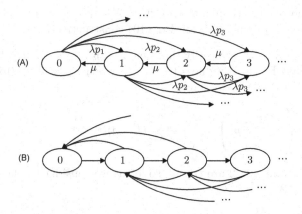

Fig. 6. (A) CTMC underlying the queue studied in Sect. 4. (B) Reversed chain.

measure of the chain. We proceed as follows. Let us consider a reference state that, without loss of generality, can be state 0. Then, given an arbitrary state i, we can compute $\pi(i)/\pi(0)$ as the ratio between the product of the rates of an arbitrary path from 0 to i divided by the product of the reversed rates of the transitions of the same path.

We are now ready to prove Theorem 2

Proof (Theorem 2). In contrast with the proof of Theorem 1, the use of time-reversed analysis does not require us to guess the expression of the stationary distribution but is entirely constructive.

In Fig. 6-(A) we show a sketch of the CTMC underlying the queue, and in Fig. 6-(B) its reversed process. Let us consider state 0: the residence time in the forward chain has rate λ. We notice that in the reversed chain, there is only one outgoing transition from state 0, i.e., that going to state 1, thus, by K1, we have: $q_{01}^R = \lambda$. If we consider the cycle $0 \to 1 \to 0$, we use K2 to obtain $q_{10}^R = \mu_1 p_1$.

The residence time in state 1 has rate $\lambda(1 - p_1) + \mu$, but now we have two outgoing transitions in the reversed chain: q_{10}^R and q_{12}^R. By K1, we can solve:

$$\lambda(1 - p_1) + \mu = q_{10}^R + q_{12}^R,$$

whose only unknown is q_{12}^R that results to be:

$$q_{12}^R = \frac{\lambda \mu p_2}{(\lambda + \mu)(1 - p_1)}.$$

Rate q_{21}^R can be derived by considering the cycle $1 \to 2 \to 1$ in the forward chain and q_{20}^R thanks to the cycle $0 \to 2 \to 1 \to 0$. In each case, we have only one unknown. It is possible to see that this holds also for the following states, and hence we can derive all the transition rates in the reversed process.

The stationary distribution is derived thanks to the relation [8] $\pi_i q_{ij} = \pi_j q_{ji}^R$. Thus, we first derive $\pi(1)$ as function of $\pi(0)$ thanks to transition $1 \to 0$, then

$\pi(2)$ as function of $\pi(1)$ thanks to transition $2 \to 1$, and so on. The regularity of the procedure allows us to obtain Eq. (10) constructively. $\quad\square$

As sanity check, it is worth to verify that Lemma 1 is a special case of Theorem 2. Let us verify that Eq. (10) reduces to Eq. (5) when $p_n = (1 - \beta)\beta^{n-1}$. Let us start with the third factor, we have:

$$\left(1 - \sum_{j=1}^{n-1} p_j\right) = \left(1 - \sum_{j=1}^{n-1}(1 - \beta)\beta^{j-1}\right) = \beta^{n-1}.$$

The cases $n = 1$ and $n = 2$ are trivial, so let us we consider the last product of Eq. (10) for $n > 2$:

$$\prod_{j=0}^{n-2}\left(\lambda\left(1 - \sum_{k=1}^{j} p_k\right) + \mu\right) = \prod_{j=0}^{n-2}(\lambda\beta^j + \mu) = \mu^{n-1}\prod_{j=0}^{n-2}\left(1 + \frac{\lambda}{\mu}\beta^j\right)$$

$$= \mu^{n-1}\left(-\frac{\lambda}{\mu}; \beta\right)_{n-1}.$$

It is now easy to see that Eq. (5) is a special case of Eq. (10).

5   Connection with Number Theory

In this Section, we show an interesting connection between the expression of $\pi(0)$ given by Eq. (5) and the two variable generating function of the triangular integer sequence identified as *A008289* [1] in the OEIS database.

Let us define $Q(n, m)$ as the number of ways that we can partition n objects in m non-empty groups in such a way that each group has a different size. For example, $Q(10, 3) = 4$, since we have:

$$10 = 1 + 7 + 2 = 1 + 6 + 3 = 1 + 5 + 4 = 2 + 5 + 3.$$

More details on this sequence can be found in [4, Ch. 2] where the following recursive relation is proved:

$$Q(n, m) = \begin{cases} 1 & \text{if } n = 1, m = 1 \\ Q(n - m, m) + Q(n - m, m - 1) & \text{if } n > m > 0 \\ 0 & \text{otherwise} \end{cases}.$$

The generating function of $Q(n, m)$ is (see, e.g., [1]):

$$g_Q(x, y) = 1 + \sum_{\substack{n>0 \\ k>0}} Q(n, k)x^n y^k = \prod_{n>0}(1 + yx^n) = \frac{\prod_{n\geq 0}(1 + yx^n)}{1 + y}$$

$$= \frac{(-y, x)_\infty}{1 + y}. \quad (11)$$

If we evaluate $g_Q(x, y)$ for $x = \beta$ and $y = \rho$ then we obtain the following elegant relation:

$$\pi(0) = \frac{1}{1 + \rho} g_Q(\beta, \rho)^{-1}.$$

From expression:

$$\sum_{n=1}^{\infty} \rho \beta^{n-1} \left(-\frac{\lambda}{\tilde{\mu}_1}; \beta \right)_{n-1}$$

we can apply the q-binomial theorem and obtain a new expression for the generating function of $Q(n, m)$:

$$g_Q(x, y) = \frac{\sum_{n=1}^{\infty} \sum_{k=1}^{\infty} x^n y^{k \frac{k+1}{2}}}{\prod_{i=1}^{k}(1 - y^i)}.$$

6   Conclusion

In this paper, we have studied a special type of queues that can host only one job consisting of a finite (but potentially unbounded) number of tasks. The queue aims at always working on the largest jobs, i.e., those jobs consisting of more tasks. A preemptive policy is adopted in case of an arrival while another job is in service. We have shown that the system admits a rather complicated explicit solution for the stationary probabilities and that this can be quickly derived thanks to a time-reversed analysis. We believe that, for this case, the guess of the stationary distribution would not be a viable route to the analysis of the queue as shown in the proof of Lemma 1. However, it would have been possible to address the problem by resorting to the method of probability generating functions that however, although more general (in its applicability) than that used for the proof of Theorem 2, poses some difficulties that are instead overcome by the time-reversed analysis. Finally, we have described some curious connections between the queueing system under analysis and the generating function of an important sequence of numbers used in combinatorial analysis.

Acknowledgements. We would like to thank prof. Michael Somos for his invaluable suggestions on the relations between the q-series considered in this paper and number theory.

References

1. The on-line encyclopedia of integer sequences. https://oeis.org/A008289. Accessed 30 Aug 2019
2. Aalto, S., Ayesta, U., Nyberg-Oksanen, E.: Two-level processor-sharing scheduling disciplines: mean delay analysis. ACM SIGMETRICS Perform. Eval. Rev. **32**(1), 97–105 (2004). Proceedings of ACM Sigmetrics/Performance
3. Balsamo, S., Harrison, P., Marin, A.: A unifying approach to product-forms in networks with finite capacity constraints. ACM SIGMETRICS Perform. Eval. Review **38**(1), 25–36 (2010). Proceedings of ACM Sigmetrics/Performance

4. Comtet, L.: Advanced Combinatorics, The Art of Finite and Infinite Expansion. D. Reidel Publishing Company, Dordrecht (1974)
5. Grosof, I., Scully, Z., Harchol-Balter, M.: SRPT for multiserver systems. Perform. Eval. **127–128**, 154–175 (2018)
6. Harrison, P.: Turning back time in Markovian process algebra. Theor. Comput. Sci. **290**(3), 1947–1986 (2003)
7. Harrison, P., Marin, A.: Product-forms in multi-way synchronizations. Comput. J. **57**(11), 1693–1710 (2014)
8. Kelly, F.P.: Reversibility and Stochastic Networks. Wiley, Hoboken (1979)
9. Marin, A., Balsamo, S., Fourneau, J.M.: LB-networks: a model for dynamic load balancing in queueing networks. Perform. Eval. **115**, 38–53 (2017)
10. Marin, A., Mitrani, I., Elahi, M., Williamson, C.: Control and optimization of the SRPT service policy by frequency scaling. In: McIver, A., Horvath, A. (eds.) QEST 2018. LNCS, vol. 11024, pp. 257–272. Springer, Cham (2018). https://doi.org/10.1007/978-3-319-99154-2_16
11. McIntosh, I.J.: Some asymptotic formulae for q-shifted factorials. Ramanujan J. **3**, 205–214 (1999)
12. Pittel, B.G.: Closed exponential networks of queues with saturation: the Jackson-type stationary distribution and its asymptotic analysis. Math. Oper. Res. **4**(4), 357–378 (1979)
13. Pradhan, S., Gupta, U.C.: Modeling and analysis of an infinite-buffer batch-arrival queue with batch-size-dependent service. Perform. Eval. **108**, 16–31 (2017)
14. Schrage, L.: A proof of the optimality of the shortest remaining processing time discipline. Oper. Res. **16**, 678–690 (1968)
15. Schroeder, B., Harchol-Balter, M.: Web servers under overload: how scheduling can help. ACM Trans. Internet Technol. **6**(1), 20–52 (2006)

Performance Evaluation of Thermal-Constrained Scheduling Strategies in Multi-core Systems

Muhammad Usama Sardar[1]([⊠])[ID], Clemens Dubslaff[2][ID],
Sascha Klüppelholz[2][ID], Christel Baier[2][ID], and Akash Kumar[1][ID]

[1] Chair for Processor Design, Technische Universität Dresden, Dresden, Germany
muhammad_usama.sardar@mailbox.tu-dresden.de, akash.kumar@tu-dresden.de
[2] Institute for Theoretical Computer Science, Technische Universität Dresden,
Dresden, Germany
{clemens.dubslaff,sascha.klueppelholz,christel.baier}@tu-dresden.de

Abstract. The increasing usage of multi-cores in safety-critical applications, such as autonomous control, demands high levels of reliability, which crucially depends on the temperature. On the other hand, there is a natural trade-off between reliability and performance. The scheduling of tasks is one of the key factors which determine the resulting system performance as well as reliability. Commonly used techniques, such as simulation based on benchmarks, can simulate only a limited number of input sequences of system runs and hardly optimize the performance-reliability trade-off. In order to accurately evaluate the schedulers and provide formal guarantees suitable in early design stages, we use formal methods for a quantitative performance-reliability trade-off analysis. Specifically, we propose to use energy-utility quantiles as a metric to evaluate the effectiveness of a given scheduler. For illustration, we evaluate TAPE, a state-of-the-art thermal-constrained scheduler, with theoretical optimal ones.

Keywords: Probabilistic model checking · Thermal-constrained scheduling · Mutli-core systems · Energy-utility quantiles

1 Introduction

The enormous increase in the processor power density [8] due to the decreasing feature size has made on-chip temperature a critical design constraint of multi-core systems. The elevated chip temperatures adversely impact other design constraints, such as reliability, performance, fault-tolerance, packaging and cooling costs [16]. High temperatures can result in more frequent transient errors and/or even permanent faults [23]. Industrial studies have demonstrated that a small difference in the operating temperature (order of 10–15 °C) can result in almost two times difference in the device lifespan [25]. Studies also show that the cooling cost increases super-linearly with the thermal dissipation [9]. Moreover, the

© Springer Nature Switzerland AG 2020
M. Gribaudo et al. (Eds.): EPEW 2019, LNCS 12039, pp. 133–147, 2020.
https://doi.org/10.1007/978-3-030-44411-2_9

static (leakage) power [26] has exponential dependence on the operating temperature, which potentially results in more thermal runaway [19]. Hence, to ensure the reliability, performance, and safety of the multi-core for modern embedded real-time systems like autonomous control [15], thermal-constrained scheduling is crucial, i.e., the thermal constraints should be accounted for in the scheduling of tasks to the cores.

Although simulation with existing benchmarks to analyze the effectiveness of the scheduling strategy is popular in the embedded system community, most of the approaches can simulate only *a limited number of input sequences* and thus may result in missing critical situations. These in turn may lead to delays in the deployment of thermal management schemes as happened in the case of Foxton thermal management that was designed for the Montecito chip [5]. It may also result in poor performance and/or thermally unsafe behaviors or even catastrophic failures at run-time, e.g., vehicle breakdown or smartphone explosion [15].

Exhaustive formal methods such as *model checking* (see, e.g., [2]) are popular for ensuring reliability of critical system parts. In this regard, a few abstract thermal models have been proposed for the formal analysis. Most of these works, e.g., [10–12,21], ignore the thermal coupling among the cores. The models without incorporating thermal coupling can result in underestimation of the temperature, which accounts for a significant difference in reliability estimation.

An application gaining more and more attention is the capability of formal methods for a quantitative trade-off analysis. *Energy-utility quantiles* [1] provide a relevant trade-off measure in probabilistic systems, e.g., systems where the environment, the workload or the occurrence of errors is modelled probabilistically. For example, a possible instance of an energy-utility quantile is the minimal number of thermal violations to ensure finishing a given number of tasks within a given time horizon with sufficiently high probability. The thermal violations refer to the situation where the temperature of a core is above the critical temperature, resulting in low functional reliability of the system. *Such properties cannot be determined using simulative approaches.*

In this paper, we present a new thermal model of multi-core systems that is simple but yet expressive enough to show realistic behaviors and analyze the model with respect to its performance-reliability trade-off properties. Particularly, our contribution comprises:

- A formal thermal model in terms of a Markov decision process (MDP, see, e.g., [2]) that, in contrast to existing works, incorporates thermal coupling of cores and transient temperatures (Sect. 2).
- A formal analysis of selected sanity checks underpinning the confidence in our abstract model (Sect. 3).
- A formal comparative performance-reliability trade-off analysis of heuristics with a scheduling strategy optimal according to the stochastic workload assumptions (Sect. 4).

2   Proposed Model

In this section, we present our proposed formal thermal model for 2D/3D multi-core systems as well as describe our workflow and parametric model. Finally, we present a model instance that we implemented for a quantitative analysis using the PRISM model checker. We denote the set of integers $i, i + 1, ..., j - 1, j$ by $[i..j]$ for $i, j \in \mathbb{N}$.

2.1   Abstract Thermal Model

For simplicity, we consider a 2D grid layout in which $N = n \times n$ homogeneous cores are placed, where we refer to a core placed at position (i, j) via an index $i + n \cdot j \in [0..N - 1]$. The model can easily be extended for 3D heterogeneous multi-core systems. Similar to the related works [3,10–12,21], we consider a discrete-time model where the power states and resulting temperatures of the cores are observed after fixed discrete intervals of time. Intuitively, the change in temperature of a core depends on three major factors: (i) power dissipated by the core, (ii) heat transferred by the core to the ambience, and (iii) heat transferred among the cores. We merge the first two factors into one called *self-heating* and model the last one as thermal coupling.

We now explain the mathematical model by equations that hold for each core with some given index i. The change in temperature of core i is given by:

$$\Delta Temp_i = w_{sh} \cdot sh(Temp_i, Power_i) + w_{cpl} \cdot cpl_i(Temp_{[0..N-1]}), \qquad (1)$$

where $\Delta Temp_i$, $Temp_i$ and $Power_i$ represent the change in temperature, the current temperature, and the current power state (e.g., ON or OFF), respectively, of core i. The weight $w_{sh} \in \mathbb{Q}$ is a constant used to represent the scaling for self-heating, which depends on the nature of the material and environmental conditions. The weight $w_{cpl} \in \mathbb{Q}$ represents the overall scaling of the thermal coupling and is specifically dependent on the conductivity of the material and the sampling interval Δt, as suggested by the general equation of heat conduction [20]. Moreover, sh and cpl represent self-heating and thermal coupling functions, as described below. The power dissipated by a core for performing some computation produces heat while the heat transferred by a core to the ambience leads to a drop in the temperature of the core. The overall effect is represented by the self-heating function sh:

$$sh(Temp_i, Power_i) = pd(Power_i) + amb(Temp_i), \qquad (2)$$

where the function pd models the increase in the temperature of a core due to the power consumed for doing some computation in one time step. For a 2-level Dynamic Voltage and Frequency Scaling (DVFS), it is defined as follows:

$$pd(Power_i) = \begin{cases} p_1, \text{ if } Power_i = 1 \\ p_2, \text{ if } Power_i = 0, \end{cases} \qquad (3)$$

where $p_1, p_2 \in \mathbb{Q}_{>0}$ are positive parameters, depending on the microarchitecture, application, and leakage characteristics. The function amb models the heat transferred to the ambience, inspired by the Newton's law of cooling [4, 20]:

$$amb(Temp_i) = -c \cdot (Temp_i - Temp_{amb}), \tag{4}$$

where $c \in \mathbb{Q}_{>0}$ is a positive parameter, depending on the cooling solution (heat sink and spreader) specifications, and $Temp_{amb}$ represents the ambient temperature. The negative sign shows that it leads to a drop in the temperature of the core.

The thermal coupling mainly depends on the conductivity of the material, sampling time, the difference of the temperature of the cores, and distance between the cores [20]. The first two factors are modeled by w_{cpl} while the last two factors vary from core to core and are captured in our model by the thermal coupling function cpl_i, as described below:

$$cpl_i(Temp_{[0..N-1]}) = \sum_{j \in [0..N-1], j \neq i} \left\{ w_{ij} \cdot (Temp_j - Temp_i) \right\}, \tag{5}$$

where while the index i still represents the index of the core under consideration, the index j in the sum represents the index of other cores. We model the weights $w_{ij} \in \mathbb{Q}$ for thermal coupling, i.e., coupling coefficients, by the reciprocal of the Euclidean distance between the two cores in the 2D grid, i.e.,

$$w_{ij} = \frac{1}{\|\boldsymbol{x} - \boldsymbol{y}\|_2^2}, \tag{6}$$

where $\boldsymbol{x}$ and $\boldsymbol{y}$ represent the 2D position vectors of the cores i and j, respectively. Considering Dirichlet's condition [22], the temperature outside the boundary of the 2D grid of cores is assumed to be the ambient temperature. Since we consider coupling coefficients w_{ij} for any pair of cores, the model is applicable to 3D multi-core systems with trivial changes.

2.2   Workflow and Parametric Model

The workflow of the proposed approach, presented in Fig. 1, begins with the thermal simulator, which requires models of floorplan, packaging and power traces. The floorplan describes sizes and placement of the cores. The user can select parameters for floorplan and packaging based on the system under consideration. Power traces of the application are then given to the thermal simulator to compute transient temperatures. The transient temperatures are then analyzed in our proposed tool to find the trends in the behavior of the temperature, which form the basis for the validation of our proposed thermal model. A reasonable continuous-valued thermal model is developed from transient temperatures. In order to find the optimal weights for Eq. (1), we use properties of symmetry with respect to power and memorylessness with respect to the initial temperature and the minimum mean square error criteria to evaluate our discretized thermal model against the continuous-valued one.

For exhaustive formal analysis, the continuous parameters, such as temperature, have to be discretized to a certain number of levels to analyze the results within a suitable time. In our parametric model, the designer can select various parameters, such as the number of temperature and power levels, number of cores in the system and the scheduling strategy (optimal/heuristic) for the analysis. In case of heuristic analysis, the scheduling criteria is required. Additionally, the designer may choose the probability distribution and its characteristics (e.g., mean) to capture the behavior of the application.

With the above parameters, we generate model variants, i.e., transition systems, Discrete-time Markov Chains (DTMCs) or MDPs, depending on the desired analysis. The purpose of having states in a transition system is the step-wise behavior induced by the task arrival and scheduling. The probability distributions model the task arrival and the non-determinism is used to model the different choices for task scheduling on the cores. Depending on the desired analysis, we provide flags in our tool for the option whether to include time or thermal violations in the state space and generate properties to be investigated accordingly. The generated formal model along with the property (e.g., quantile query) is input to the probabilistic model checker to perform the analysis (e.g., to compute quantiles for various probability thresholds). The output logs from the probabilistic model checker are input to our tool to generate quantile plots as well as analyze model sizes. Thus, our approach helps the designer to perform performance-reliability trade-off analysis for the designed scheduler even with minimal prior knowledge of formal methods.

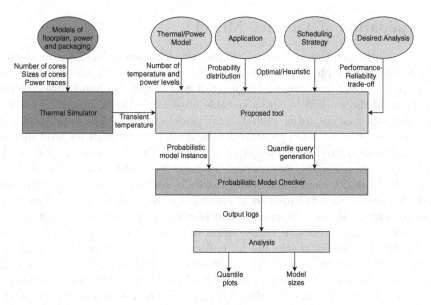

Fig. 1. Workflow of the proposed approach

We use a popular thermal simulator, HotSpot [24], for the thermal analysis. Since MATLAB is one of the most commonly used software environment for performance analysis, we utilize MATLAB to develop our tool. For probabilistic model checking, we choose PRISM [14] as it supports computing quantiles.

2.3 Concrete Model

For illustrative purposes, we present the details of the concrete model, instantiated from the parametric model presented above, for the configurations used in our experiment: the system has 9 cores arranged in a 2D grid of 3×3. We chose a granularity of 3 temperature levels, represented as 0, 1 and 2, respectively, for our experiments. We further considered 2 level DVFS, i.e., core power has 2 states (ON or OFF).

Further parameters chosen are $p_1 = 1$ and $p_2 = 0$ in Eq. (3), i.e., the temperature rise due to power dissipation in one time step is unity and zero, when the core is powered ON and OFF, respectively. Moreover, the value of constant c for heat sink and spreader specifications in Eq. (4) is selected as 0.25. For brevity, in this work we consider thermal coupling effects from 4 direct neighbors of each core. The cores are considered to be unit distance apart, so that the weights w_{ij} in Eq. (6) are taken as unity for all the 4 direct neighbors. The selected weights for Eq. (1) for the above parameters by using the minimum mean square error are 1.3 for self-heating (w_{sh}) and 0.09 per core per temperature level difference for thermal coupling (w_{cpl}). It should be noted that the concrete model is a proof-of-concept and we do not claim to share the realistic parameters.

3   Validation of Thermal Model

While our abstract thermal model we presented in Sect. 2 has been motivated by principles from physics, the focus of the work is neither to find an exact model nor a model reflecting every aspect of heat transfer in multi-core systems. The focus of the work is to find a suitable implementation that can be subject of a trade-off analysis using probabilistic model checking. The simplifications, e.g., discretization to 3 temperature levels, done towards such an implementation could have introduced severe side-effects. Hence, to increase confidence in the methods we proposed, we analyzed the concrete model presented in Sect. 2.3 against basic sanity checks one would naturally assume to hold in any thermal model for heat transfer of multi-core systems. In the initial configuration, there is no constraint on the temperature and power of any core, i.e., all initial configurations with any possible combination of temperature and power levels are considered. The sanity checks are formalized in Linear Temporal Logic (LTL) [2], with temporal modalities *eventually*, *always*, and *next* by $\Diamond$, $\Box$, and $\bigcirc$, respectively. We first considered the standard model with full non-determinism in the power level switches and then a variant with non-determinism only in the initial configurations, i.e., power levels cannot be switched anymore after the initial time-step. For the actual analysis, we used the probabilistic model checker PRISM [14] with its support of verifying (non-probabilistic) LTL properties.

3.1   Non-restricted Power Level Switches

First, we considered persistence properties (cf., e.g., [2]) that should be fulfilled in any scenario of powering cores in the multi-core system. Our proposed thermal model satisfies the following properties for all the 9 cores starting from any possible initial state (any power or temperature level for any core):

1. *Non-decreasing temperature*: Whenever a core is turned ON and its current temperature is less than or equal to the steady-state temperature, the temperature of the core at the next instant is greater than or equal to its current temperature. The second condition is required because if the temperature of the core is above the steady-state temperature, it may decrease even if the core is powered ON because of the thermal coupling effect. For each core i, this is formalized by the LTL formula:

$$\Box((Power_i = ON \wedge Temp_i \leqslant Temp_{SSi}) \Rightarrow (\bigcirc Temp_i \geqslant Temp_i)), \quad (7)$$

where $\wedge$ represents the conjunction of the predicates and $Power_i$ represents the power status of the core i. Moreover, $Temp_{SSi}$ represents the steady state temperature of a core i and $\bigcirc Temp_i$ represents the temperature of the core i at the next instant.

2. *Non-increasing temperature*: Whenever a core is turned OFF and its current temperature is greater than the minimum (ambient) temperature, the temperature of the core at the next instant is less than or equal to its current temperature. The second condition is required because if the core is at the ambient temperature, it may increase even if the core is powered OFF because of the thermal coupling effect. For each core i, this is formalized by an LTL formula:

$$\Box((Power_i = OFF \wedge Temp_i > Temp_{amb}) \Rightarrow (\bigcirc Temp_i \leqslant Temp_i)), \quad (8)$$

where $Temp_{amb}$ represents the ambient temperature.

3.2   Fixed Power Levels

While in the last section, power levels could change nondeterministically over time, we now consider properties regarding limit behaviors when fixing power levels (voltage and frequency) on cores. The reason for fixing power levels is that such properties could be observed using the temperature data from HotSpot. We could show that our proposed thermal model satisfies the following properties starting from any possible initial state (any temperature level for any core):

1. *Drop to initial temperature at power-OFF*: If from some point onwards all cores remain powered OFF, the temperatures of all cores eventually drop to the initial temperature and stay there. In HotSpot, this case was analyzed by giving a power trace with each core consuming 0 m W after some point in time. Formally, in every path, if at some point of time, all cores remain continuously

powered OFF, then from some moment on all cores remain continuously at the initial temperature. This is formalized as follows:

$$\Diamond(\, \Box \wedge_{0 \leq i \leq 8}(Power_i = OFF) \Rightarrow \Diamond \Box \wedge_{0 \leq i \leq 8}(Temp_i = Temp_{init})), \quad (9)$$

where $Temp_{init}$ represents the initial temperature. For the analysis of this property, all power levels are initially set to OFF in the PRISM model and remain OFF throughout. Our proposed model satisfies this property because all cores disseminate heat to the ambience. Also, the hotter cores continue to transfer heat to the cooler cores to balance the temperature. So from any starting temperature, the cores eventually end up in the initial temperature.

2. *Maximum temperature at the central core*: If from some point onwards all cores remain powered ON, the temperature of the central core eventually remains the maximum of all cores in the 3×3 core system. In HotSpot, this was tested using same power dissipation for all the cores. This is formalized as follows:

$$\Diamond(\, \Box \wedge_{0 \leq i \leq 8}(Power_i = ON) \Rightarrow \Diamond \Box \, (Temp_4 = max(Temp_0, ..., Temp_8))).$$
$$(10)$$

For the analysis of this property, all the power levels are initially set to ON in the PRISM model.

3. *Average temperature property*: If from some point onwards all cores remain powered ON, the temperature of the central core in the 3×3 core system eventually remains greater than or equal to the average temperature of its 4 neighboring cores. In HotSpot, this was tested using same power dissipation for all the cores. This is formalized as follows:

$$\Diamond(\, \Box \wedge_{0 \leq i \leq 8}(Power_i = ON) \Rightarrow \Diamond \Box \, (Temp_4 \geq \frac{1}{4} \cdot \sum_{j \in Neigh} (Temp_j))), \quad (11)$$

where $Temp_4$ represents the temperature of the central core and $Neigh$ represents the set of indices of direct neighbors of the central core, i.e., $\{1,3,5,7\}$. For the analysis of this property, all the power levels are initially set to ON in the PRISM model.

4   Comparative Analysis of Heuristics

In this section, we present a refined formal model that reflects the system behavior in practical applications, with additional stochastic information, e.g., on the workload of the system. For including thermal management strategies in our model to distribute workload on the cores, we add non-deterministic choices of the cores to be powered ON as soon as the required amount of workload is apparent. This yields a MDP with probabilistic workload choices and non-deterministic powering of cores. Our model then does not only pave the way for a best- and worst-case analysis, revealing optimal thermal management policies for the assumed probabilistic workload profile, but can also be used to analyze existing thermal management heuristics.

4.1 Formal Model

Inspired by the work [18] from the literature, we consider that the number of tasks arriving in the multi-core at each time instant follows the Poisson distribution. For a N-core system, we truncate and normalize the Poisson distribution for $0 - N$ tasks. We compute the probabilities in our MATLAB-based tool and export to PRISM as constants. Similar to simulation-based analysis [17], we assume that there is no data dependency among tasks, i.e., we consider independent tasks. For brevity, each task is assumed to have an execution time of one time step. Moreover, we assume that each core executes only one task at a time.

For the optimal scheduler, we use non-determinism to capture the possible ways that a controller can influence the behaviour of the system. Since we are interested in determining which cores should be turned ON, the non-determinism is in the selection of a core to run a task and the decision whether to put the tasks in a queue. The optimal scheduler is then computed that resolves all the non-deterministic choices such that the expected values are either maximized or minimized.

For the PRISM implementation, the task mapping is implemented in two phases, utilizing different synchronization labels. In the first phase, all cores are powered OFF and the number of tasks is selected probabilistically depending on the parameterized mean value of the Poisson distribution. The temperature from the previous step is also updated in the first phase. Then, in the second phase, a core is non-deterministically selected from the cores which are currently OFF for the task to be mapped and the number of tasks is decremented by one. The process is repeated until the number of tasks becomes zero, or the number of tasks left to be mapped are less than or equal to the queue size and no more cores are thermally suitable for mapping the tasks (i.e., the temperature of all available cores is greater than or equal to the threshold temperature). Then, the two phases are repeated. In the initial configuration, all cores are turned OFF, all cores are at the ambient temperature, the queue is empty and there are no tasks to be mapped to the multi-core. In order to analyze the temperature, time, number of thermal violations and tasks, transition rewards [7] are utilized. The rewards are updated on the transition to the selection of new tasks.

For the heuristics, the non-determinism is resolved by the specific schedulers, resulting in a DTMC [2]. In this work, we consider 3 specific schedulers with a defined mapping criteria. For more than 1 core satisfying the mapping criteria, to resolve the non-determinism in each case, we use a specific order, i.e., [0 8 6 2 1 3 5 7 4], where the numbers indicate the indices of the cores, e.g., core 8 is scheduled before core 5. The mapping criteria of considered heuristics are presented below: A popular thermal-aware scheduler, TAPE [6], is based on the economic model and maps the tasks to the core with the criteria $max(sellT_i - buyT_i)$, and in case of multiple cores satisfying this criteria: $min(abs(buyT_i))$, where $sellT_i$ and $buyT_i$ represents *sell* and *buy* values, respectively, of a core i at temperature T_i. For modelling TAPE, we use the same weights as presented in the paper [6]. The reactive thermal-aware schedulers map to the cores with the criteria

$Temp_i < TempThreshold$ and the minimum temperature scheme maps to the coolest core among the currently available cores, i.e., $min(Temp_i)$.

4.2   Comparative Trade-Off Analysis

The formal model presented above is parameterized in the queue size and mean value. For a real-world scenario, the queue size is selected according to the system under consideration and the mean value is selected based on the prior information about the workload. In the following analysis, we consider the mean value of 7.5 and queue size of 3.

We propose to use performance metrics for the evaluation of a thermal-aware scheduler based on *energy-utility quantiles* [1]. Within energy-utility quantiles, two reward structures formalize quantities of the system and a trade-off condition is posed by putting bounds on the accumulated reward during an execution. Varying one of the bounds and optimizing this value such that the probability mass of paths with the accumulated rewards staying within the bounds exceeds a given threshold provides a trade-off metric that can be computed using probabilistic model checking [1].

Maximal Time to Thermal Violations. One of the most important functionalities of a thermal-aware scheduler is to maximize the system's thermal stability, in terms of the time to a certain number of violations. So, we consider the following energy-utility quantile: what is the maximal time the system survives with probability p until reaching a certain number of thermal violations for some scheduler. Formally, this can be expressed as the following *existential energy-utility upper-bound quantile* (path formula is increasing and state property is decreasing) [1]:

$$max \left\{ t : Pr_s^{min}\left(\Diamond \left(\textit{Time} \leqslant t \ \wedge \ \# \textit{Violations} \geqslant v\right)\right) \leqslant p \right\}, \qquad (12)$$

where $v \in \mathbb{N}$ represents a lower bound on the accumulated number of global thermal violations, $t \in \mathbb{N}$ represents an upper bound on time and $p \in [0,1] \cap \mathbb{Q}$ represents the probability threshold. The results for $v = 30$ are presented in Fig. 2 and show that TAPE and minimum temperature heuristic have the same trade-off characteristics. This is because task migration is not considered in this work. Both of them perform better than the reactive heuristic.

Minimal Thermal Violations in a Specific Time. The number of thermal violations in a specific time is an indication of the lifetime reliability of the system. So, we consider the following quantile: the minimal number of thermal violations in a specific time with at least a probability of p. Formally, this can be expressed as the following *existential energy-utility upper-bound quantile* (path formula is increasing and state property is increasing) [1]:

$$min \left\{ v : Pr_s^{max}\left(\Diamond \left(\# \textit{Violations} \leqslant v \ \wedge \ \textit{Time} = t\right)\right) > p \right\}, \qquad (13)$$

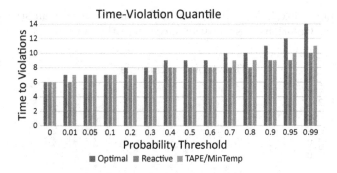

Fig. 2. Time to reach 30 global thermal violations

where $v \in \mathbb{N}$ represents a lower bound on the accumulated number of global thermal violations, $t \in \mathbb{N}$ represents the time for analysis and $p \in [0,1] \cap \mathbb{Q}$ represents the probability threshold. The results for $t = 10$ are presented in Fig. 3. Other than the low probability thresholds (covering only a few practical cases), the heuristics are near-optimal on this criteria. For instance, at a probability threshold of 0.99, the heuristics exhibit a single thermal violation more than the optimal scheduler. So, there is less room for improvement in this case.

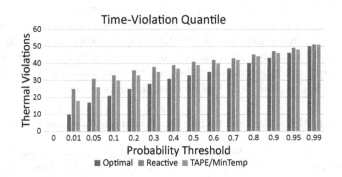

Fig. 3. Number of global thermal violations in 10 time steps

Minimal Consecutive Thermal Violations in a Specific Time. From the perspective of core lifetime reliability, it is important how long the thermal violation stays on a core. In this regard, we compute the quantile, similar to (13):

$$min \left\{ c : Pr_s^{max}\left(\Diamond(\# Consec Vio \leqslant c \wedge Time = t) \right) > p \right\}, \qquad (14)$$

where c represents the number of consecutive thermal violations. The results for $t = 10$ are presented in Fig. 4 and conforms with the above findings. Specifically, at a probability threshold of 0.99, the TAPE and reactive heuristics exhibit 1 and 2 thermal violations, respectively, more than the optimal scheduler.

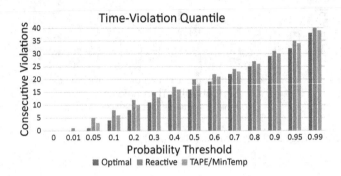

Fig. 4. Number of global consecutive thermal violations in 10 time steps

The model sizes and timings are presented in Table 1. The model size is represented in terms of the number of states, transitions and binary decision diagram (BDD) nodes. The heuristics are small enough to be analyzed by the faster explicit engine in the PRISM model checker. Hence, BDD nodes are not provided. The timings for building the model and computation of quantiles are also provided. For instance, the computation of the optimal scheduler of quantile (14) takes a couple of days. For such model sizes, the computation of advanced properties like quantiles may take a few days [13]. However, various experiments were performed to ensure the scalability of the approach with various temperature levels and quantile bounds. The results show that the time for model checking does not explode with the increase in temperature levels.

Table 1. Model sizes and timings for optimal scheduler and heuristics

			Model sizes			Timings [s]	
Quantile	Scheduler	Constants	States	Transitions	BDD nodes	Construct	Compute
(12)	Optimal	v = 30	1809781760	10738895130	25134175	622	50310
	Reactive		6940244	19164152	-	492	538
	TAPE		15128017	41888689	-	2666	2815
	MinTemp		15128017	41888689	-	734	2697
(13)	Optimal	t = 10	465432065	2762915430	4954803	118	65605
	Reactive		1206986	3320915	-	102	24
	TAPE		2137295	5901950	-	261	38
	MinTemp		2137295	5901950	-	149	41
(14)	Optimal	t = 10	465432065	2762915430	4954803	103	156566
	Reactive		1206986	3320915	-	69	32
	TAPE		2137295	5901950	-	226	69
	MinTemp		2137295	5901950	-	195	87

5   Conclusions

In this work, we presented a formal thermal model for multi-cores incorporating thermal coupling as well as transient temperatures. This is challenging because HotSpot gives only the final temperature instead of the individual components of self-heating and thermal coupling. The presented model is validated against various sanity checks for non-restricted power level switches as well as constant power levels. We proposed a quantitative performance-reliability trade-off analysis, based on quantiles, of thermal-aware scheduling strategies for multi-core systems. The results show that the evaluated scheduler TAPE can be improved with respect to the maximal time to reach a certain number of thermal violations. Our approach thus helps in the evaluation of heuristics. For future, the evaluation of heuristics for heterogeneous multi-core systems can be very interesting. Moreover, performance evaluation in terms of throughput of a scheduler can also be interesting.

Acknowledgments. We would like to thank Steffen Märcker for debugging memory related issues in the quantile implementation of PRISM.

Christel Baier, Clemens Dubslaff, and Sascha Klüppelholz are supported by the DFG through the Collaborative Research Centers CRC 912 (HAEC) and TRR 248 (see https://perspicuous-computing.science, project ID 389792660), the Clusters of Excellence EXC 2050/1 (CeTI, project ID 390696704) and EXC 1056 (cfAED) as part of Germany's Excellence Strategy, and the Research Training Groups QuantLA (GRK 1763) and RoSI (GRK 1907). Akash Kumar is supported by the DFG through the Cluster of Excellence EXC 1056 (cfAED).

References

1. Baier, C., Daum, M., Dubslaff, C., Klein, J., Klüppelholz, S.: Energy-utility quantiles. In: Badger, J.M., Rozier, K.Y. (eds.) NFM 2014. LNCS, vol. 8430, pp. 285–299. Springer, Cham (2014). https://doi.org/10.1007/978-3-319-06200-6_24
2. Baier, C., Katoen, J.P.: Principles of Model Checking. MIT Press, Cambridge (2008)
3. Bukhari, S.A.A., Lodhi, F.K., Hasan, O., Shafique, M., Henkel, J.: FAMe-TM: formal analysis methodology for task migration algorithms in many-core systems. Sci. Comput. Program. **133**(2), 154–174 (2017). https://doi.org/10.1016/j.scico.2016.06.004
4. Burmeister, L.C.: Convective Heat Transfer. Wiley, Hoboken (1993)
5. Dunn, D.: Intel delays Montecito in roadmap shakeup. EE Times, Manufacturing/Packaging (2005)
6. Ebi, T., Al Faruque, M.A., Henkel, J.: TAPE: thermal-aware agent-based power economy multi/many-core architectures. In: Computer-Aided Design, pp. 302–309. IEEE (2009). https://doi.org/10.1145/1687399.1687457
7. Forejt, V., Kwiatkowska, M., Norman, G., Parker, D.: Automated verification techniques for probabilistic systems. In: Bernardo, M., Issarny, V. (eds.) SFM 2011. LNCS, vol. 6659, pp. 53–113. Springer, Heidelberg (2011). https://doi.org/10.1007/978-3-642-21455-4_3

8. Gnad, D., Shafique, M., Kriebel, F., Rehman, S., Sun, D., Henkel, J.: Hayat: harnessing dark silicon and variability for aging deceleration and balancing. In: Design Automation Conference, pp. 1–6. ACM/EDAC/IEEE (2015). https://doi.org/10.1145/2744769.2744849

9. Gunther, S.H., Binns, F., Carmean, D.M., Hall, J.C.: Managing the impact of increasing microprocessor power consumption. Intel Technol. J. 1–9 (2001)

10. Iqtedar, S., Hasan, O., Shafique, M., Henkel, J.: Formal probabilistic analysis of distributed dynamic thermal management. In: Design, Automation and Test in Europe, pp. 1221–1224. IEEE (2015). https://doi.org/10.7873/DATE.2015.0503

11. Iqtedar, S., Hasan, O., Shafique, M., Henkel, J.: Probabilistic formal verification methodology for decentralized thermal management in on-chip systems. In: Enabling Technologies: Infrastructures for Collaborative Enterprises, pp. 210–215. IEEE (2015). https://doi.org/10.1109/WETICE.2015.39

12. Ismail, M., Hasan, O., Ebi, T., Shafique, M., Henkel, J.: Formal verification of distributed dynamic thermal management. In: Computer-Aided Design, pp. 248–255. IEEE (2013). https://doi.org/10.1109/ICCAD.2013.6691126

13. Klein, J., et al.: Advances in probabilistic model checking with PRISM: variable reordering, quantiles and weak deterministic Büchi automata. Int. J. Softw. Tools Technol. Transfer **20**(2), 179–194 (2017). https://doi.org/10.1007/s10009-017-0456-3

14. Kwiatkowska, M., Norman, G., Parker, D.: PRISM 4.0: verification of probabilistic real-time systems. In: Gopalakrishnan, G., Qadeer, S. (eds.) CAV 2011. LNCS, vol. 6806, pp. 585–591. Springer, Heidelberg (2011). https://doi.org/10.1007/978-3-642-22110-1_47

15. Lee, Y., Chwa, H.S., Shin, K.G., Wang, S.: Thermal-aware resource management for embedded real-time systems. Comput.-Aided Des. Integr. Circuits Syst. **37**(11), 2857–2868 (2018). https://doi.org/10.1109/TCAD.2018.2857279

16. Liu, Z., Tan, S.X.D., Huang, X., Wang, H.: Task migrations for distributed thermal management considering transient effects. IEEE Trans. Very Large Scale Integr. Syst. **23**(2), 397–401 (2015). https://doi.org/10.1109/TVLSI.2014.2309331

17. Pagani, S., Chen, J., Shafique, M., Henkel, J.: Advanced Techniques for Power, Energy, and Thermal Management for Clustered Manycores. Springer, Cham (2018). https://doi.org/10.1007/978-3-319-77479-4

18. Pathania, A., Venkataramani, V., Shafique, M., Mitra, T., Henkel, J.: Defragmentation of tasks in many-core architecture. Archit. Code Optim. **14**(1), 2:1–2:21 (2017). https://doi.org/10.1145/3050437

19. Pedram, M., Nazarian, S.: Thermal modeling, analysis, and management in VLSI circuits: principles and methods. Proc. IEEE **94**(8), 1487–1501 (2006). https://doi.org/10.1109/JPROC.2006.879797

20. Remsburg, R.: Advanced Thermal Design of Electronic Equipment. Springer, Heidelberg (2011)

21. Sardar, M.U., Hasan, O., Shafique, M., Henkel, J.: Theorem proving based formal verification of distributed dynamic thermal management schemes. J. Parallel Distrib. Comput. **100**, 157–171 (2017). https://doi.org/10.1016/j.jpdc.2016.06.011

22. Sbalzarini, I.F.: Spatiotemporal modeling and simulation (2016)

23. Srinivasan, J., Adve, S.V., Bose, P., Rivers, J.A.: Lifetime reliability: toward an architectural solution. IEEE Micro **25**(3), 70–80 (2005). https://doi.org/10.1109/MM.2005.54

24. Stan, M.R., Skadron, K., Barcella, M., Huang, W., Sankaranarayanan, K., Velusamy, S.: HotSpot: a dynamic compact thermal model at the processor-architecture level. Microelectronics **34**(12), 1153–1165 (2003). https://doi.org/10.1016/S0026-2692(03)00206-4

25. Viswanath, R., Wakharkar, V., Watwe, A., Lebonheur, V.: Thermal performance challenges from silicon to systems. Intel Technol. J. 1–16 (2000)

26. Yeo, I., Liu, C.C., Kim, E.J.: Predictive dynamic thermal management for multicore systems. In: Design Automation Conference, pp. 734–739. ACM/IEEE (2008). https://doi.org/10.1145/1391469.1391658

Bounding the Rate of Convergence for One Class of Finite Capacity Time Varying Markov Queues

Alexander Zeifman[1] , Yacov Satin[2] , Rostislav Razumchik[3(✉)] ,
Anastasia Kryukova[2], and Galina Shilova[2]

[1] Vologda State University, Institute of Informatics Problems,
Federal Research Center "Computer Science and Control" of the Russian Academy
of Sciences, Vologda Research Center of RAS, Vologda, Russia
a_zeifman@mail.ru
[2] Vologda State University, Vologda, Russia
krukovanastya25@mail.ru, yacovi@mail.ru, shgn@mail.ru
[3] Institute of Informatics Problems, Federal Research Center
"Computer Science and Control" of the Russian Academy of Sciences,
Peoples' Friendship University of Russia (RUDN University), Moscow, Russia
rrazumchik@ipiran.ru, razumchik-rv@rudn.ru

Abstract. Consideration is given to the two finite capacity time varying Markov queues: the analogue of the well-known time varying $M/M/S/0$ queue with S servers each working at rate $\mu(t)$, no waiting line, but with the arrivals happening at rate $\lambda(t)$ only in batches of size 2; the analogue of another well-known time varying $M/M/1/(S-1)$ queue, but with the server, providing service at rate $\mu(t)$ if and only if there are at least 2 customers in the system, and with the arrivals happening only in batches of size 2. The functions $\lambda(t)$ and $\mu(t)$ are assumed to be non-random non-negative analytic functions of t. The new approach for the computation of the upper bound for the rate of convergence is proposed. It is based on the differential inequalities for the reduced forward Kolmogorov system of differential equations. Feasibility of the approach is demonstrated by the numerical example.

Keywords: Queueing systems · Rate of convergence ·
Non-stationary · Markovian queueing models · Limiting characteristics

1 Introduction

Non-stationary Markovian queueing models have been actively studied over the past few decades (see [1–6,9,15,16,19,21] and the references therein) and the interest in this topic seems not to be declining. There exists one (to some extent)

This research was supported by Russian Science Foundation under grant 19-11-00020.

M. Gribaudo et al. (Eds.): EPEW 2019, LNCS 12039, pp. 148–159, 2020.
https://doi.org/10.1007/978-3-030-44411-2_10

general framework for the analysis of such systems, which was developed in the series of papers by the authors. It consists of the following four steps[1]:

(a) find the upper bounds for the rate of convergence to the limiting regime[2];
(b) find the lower bounds for the rate of convergence to the limiting regime, which demonstrate that the dependence on the initial condition cannot vanish before some time instant t_*;
(c) obtain the stability (perturbation) bounds providing that if the structure of the rate (generator) matrix of the process is taken into account in an appropriate way, and the errors in the transition rates are small, then the basic characteristics of the process are calculated in an adequate way;
(d) approximate the process $X(t)$ by a similar, but truncated processes with a smaller number of states and construct the corresponding estimates for the approximation error.

By carrying out the steps (a)—(d) for the system with 1−time-periodic rates and by solving the forward Kolmogorov system of differential equations (like (6)) with the simplest initial condition $X(0) = 0$ for the truncated process on the interval $[t^*, t^* + 1]$ one obtains all basic probability characteristics of the process $X(t)$ and the "perturbed" processes. It is worth noticing that the step (a) is the most important among the four. This is due to the fact that once the upper bounds are obtained all other steps (b)—(d) can be carried out in a straightforward manner, by using the results of [19–25].

It is worth noticing that exact estimates of the rate of convergence yield exact estimates of stability (perturbation) bounds (see [7,8,10–13,17,20] and references therein).

In the previous two papers [14,27] one has outlined the new approach for the computation of the upper bound for the rate of convergence, which is based on the application of differential inequalities to the reduced forward Kolmogorov system of differential equations. Here one presents the detailed description of this approach for the case of finite state inhomogeneous Markov chains (see Sect. 2). Its feasibility is demonstrated on one class of non-stationary Markovian queues (see Sect. 3). In the Sect. 4 the numerical example is given. Section 5 concludes the paper.

Throughout the paper vectors are regarded as column vectors, ^{T} denotes the matrix transpose. The norm of a vector is denoted by $\| \cdot \|$ and means the

---

[1] For the more detailed description the reader is referred to [26].

[2] The limiting regime implies that beginning from a certain time instant, say, t^*, the probability characteristics of the process $X(t)$ for $t > t^*$ do not depend on the initial conditions (up to a given discrepancy). Note that a Markov chain $X(t)$ is called weakly ergodic, if $\|\boldsymbol{p}^*(t) - \boldsymbol{p}^{**}(t)\| \to 0$ as $t \to \infty$ for any initial conditions $\boldsymbol{p}^*(0)$ and $\boldsymbol{p}^{**}(0)$, where $\boldsymbol{p}^*(t)$ and $\boldsymbol{p}^{**}(t)$ are the corresponding solutions of (6). When considering weak ergodicity and inhomogeneous Markov chains, in general, any regime may be regarded as a limiting one. For example, in the case when the transition rates are 1−periodic functions, the system (6) has 1−periodic solution in the weak ergodic sense and it is reasonable to regard this solution as limiting.

sum of the absolute values of the vector's elements. When a vector, say $\boldsymbol{x}(t)$, is considered only for t from the fixed interval, say I, and not from the whole real positive line, the notation $\|\boldsymbol{x}(t)\|_I$ is used.

2 Description of the Approach

Consider a homogeneous system of linear differential equations in the vector-matrix form:

$$\frac{d}{dt}\boldsymbol{x}(t) = K(t)\boldsymbol{x}(t), \tag{1}$$

where $\boldsymbol{x}(t)$ is the real column vector and $K(t)$ is the $S \times S$ matrix with the elements $k_{ij}(t)$, being real functions, which are analytic for any $t \geq 0$. Let $\boldsymbol{x}(t)$ be the non-trivial solution of (1). Fix an arbitrary time instant $t = t_0$. Assume for now that $x_1(t_0) > 0$. Due to the continuity assumption for some value $\epsilon_1 > 0$ $x_1(t_0)$ remains positive in the interval $I_1 = (t_0 - \epsilon_1, t_0 + \epsilon_1)$. For other $S - 1$ elements of $\boldsymbol{x}(t_0)$ one can find other appropriate intervals $I_2, \dots, I_S$ in which the sign of the corresponding element does not change. Denote by $I = (t_1, t_2)$ the intersection of all of these intervals i.e. $I = I_1 \cap \dots \cap I_S$. In this common interval I the signs of the elements of $\boldsymbol{x}(t)$ do not change. Let us assume that $x_i(t) < 0$ for $i \in \{i_1, \dots, i_k\} \subset \{1, \dots, S\}$ and $x_i(t) \geq 0$ for $i \in \{1, \dots, S\} \backslash \{i_1, \dots, i_k\}$. Choose S positive numbers, say $\{d_1^I, \dots, d_S^I\}$, and put $z_i(t) = -d_i^I x_i$ if $i \in \{i_1, \dots, i_k\}$ and $z_i(t) = d_i^I x_i$ otherwise. Then $z_i(t) \geq 0$ for all $t \in I$ and $i \in \{1, \dots, S\}$ and thus $\sum_{i=1}^S z_i(t)$ is the norm of the vector $\boldsymbol{z}(t)$ in the interval I. By differentiating $\|\boldsymbol{z}(t)\|_I$ with respect to t, one has:

$$\frac{d}{dt}\|\boldsymbol{z}(t)\|_I = \sum_{i=1}^S \frac{dz_i(t)}{dt} = \sum_{j=1}^S \sum_{i=1}^S \underbrace{\frac{d_i^I}{d_j^I}\vartheta_{ij}k_{ij}(t)}\,z_j(t) = \sum_{j=1}^S \alpha_j^I(t)z_j(t), \tag{2}$$
$$\alpha_j^I(t)$$

where $\vartheta_{ij} = 1$ if $x_i(t)$ and $x_j(t)$ are of the same sign and $\vartheta_{ij} = -1$ otherwise. Therefore from (2) one has the following upper bound

$$\frac{d}{dt}\|\boldsymbol{z}(t)\|_I \leq \alpha^I(t)\|\boldsymbol{z}(t)\|_I, \tag{3}$$

where $\alpha^I(t) = \max_{1 \leq j \leq S} \alpha_j^I(t)$ and thus

$$\|\boldsymbol{z}(\tau_2)\|_I \leq e^{\int_{\tau_1}^{\tau_2} \alpha^I(u)\,du}\|\boldsymbol{z}(\tau_1)\|_I,$$

for any $t_1 \leq \tau_1 \leq \tau_2 \leq t_2$. By comparing the norms $\|\boldsymbol{z}(t)\|_I$ and $\|\boldsymbol{x}(t)\|$ one obtains the following upper bound for the $\|\boldsymbol{x}(t)\|$:

$$\|\boldsymbol{x}(\tau_2)\| \leq C^I e^{\int_{\tau_1}^{\tau_2} \alpha^I(u)\,du}\|\boldsymbol{x}(\tau_1)\|, \tag{4}$$

for any $t_1 \leq \tau_1 \leq \tau_2 \leq t_2$, where $C^I = \frac{\max_{1 \leq i \leq S} d_i^I}{\min_{1 \leq i \leq S} d_i^I}$.

Note that the first step in the derivation of (4) was the assumption that some elements of $x(t)$ are negative and the other are non-negative in I. But since the total number of elements in $x(t)$ is S there are a total of 2^S such assumptions (i.e. 2^S possible combinations of elements' signs in $x(t)$). Let us assume that for each of the 2^S combinations one can find proper I and $\{d_i^I, 1 \leq i \leq S\}$, and thereby compute $\alpha^I(t)$ and C^I. Thus one has 2^S upper bounds of type (4) and among them one can choose the worst one. Note that if for some t the two-sided derivative of $\|x(t)\|$ does not exist, it can be replaced by the right-hand derivative. Thereby all possible combinations of elements' signs in $x(t)$ are considered and the following theorem holds.

Theorem. *Let all $k_{ij}(t)$ be analytic functions of t for $t \geq 0$. Then for any $0 \leq s \leq t$ and any initial condition $\|x(s)\|$ the following bound holds:*

$$\|x(t)\| \leq C e^{\int_s^t \alpha(u)\, du} \|x(s)\|, \tag{5}$$

where $C = \max_{\text{all } I} C^I$, $\alpha(t) = \max_{\text{all } I} \alpha^I(t)$.

In the next section it is being demonstrated how this approach works in the case of several Markov queues with time varying arrival and service rates.

3 Model Description

Consider[3] a time varying $M/M/\cdot/S$ queue in which customers arrive only in the batches of size 2 with rate $\lambda(t)$. If a pair of customers arrives but there is no free room in the system for both customers, they both are lost. The service rate may depend on the total number of customers in the system and is equal to $\mu_i(t)$, when i customers are present in the system. Clearly, $\mu_0(t) = 0$. The functions $\lambda(t)$ and $\mu_i(t)$ are assumed to be non-random non-negative analytic functions of t.

In the notation $M/M/\cdot/S$ one has not specified the number of servers in the system. This is due to the fact that the number of servers explicitly depends on the values of $\mu_i(t)$. In what follows two extreme cases are considered:

(i) $\mu_i(t) = i\mu(t)$, $1 \leq i \leq S$, which means that the considered queue is the analogue of the well-known time varying $M/M/S/0$ queue with S servers each working at rate $\mu(t)$, no waiting line, but with the arrivals happening only in the batches of size 2;

(ii) $\mu_1(t) = 0$ and $\mu_i(t) = \mu(t)$, $2 \leq i \leq S$, which means that the considered queue is the variant of another well-known time varying $M/M/1/(S-1)$ queue, but this time with the server, providing service (at rate $\mu(t)$) if and only if there are at least 2 customers in the system, and with the arrivals happening only in the batches of size 2. Note that here only one customer at a time may be served.

---

[3] This is one of the four classes of systems considered in [24, 25].

For the time being it is more convenient to assume that the service rate in the system is equal to $\mu_i(t)$ and do not specify which of the two cases, (i) or (ii), is being considered.

Let $X(t)$ be the Markov process, equal to the total number of customers in the system at time t i.e. $X(t)$ takes values in the finite set $\mathcal{X} = \{0, 1, \ldots, S\}$. Denote by $p_{ij}(s, t) = P\{X(t) = j \,|\, X(s) = i\}$, $i, j \geq 0$, $0 \leq s \leq t$, the transition probabilities of $X(t)$ and by $p_i(t) = P\{X(t) = i\}$—the probability that $X(t)$ is in state i at time t. Let $\boldsymbol{p}(t) = (p_0(t), p_1(t), \ldots, p_S(t))^{\mathrm{T}}$ be probability distribution vector at instant t. Throughout the paper it is assumed that in a small time interval h the possible transitions and their associated probabilities are

$$
p_{ij}(t, t+h) = \begin{cases} q_{ij}(t)h + \alpha_{ij}(t, h), & \text{if } j \neq i, \\ 1 - \sum\limits_{k \in \mathcal{X}, k \neq i} q_{ik}(t)h + \alpha_i(t, h), & \text{if } j = i, \end{cases}
$$

where $q_{ij}(t)$ are the transition rates and $\alpha_{ij}(t, h) = o(h)$ for all i, j. For the queueing system under consideration the transition rates can be easily specified: $q_{i,i+2}(t) = \lambda(t)$, $0 \leq i \leq S - 2$, and $q_{i,i-1}(t) = \mu_i(t)$, $1 \leq i \leq S$.

The vector $\boldsymbol{p}(t)$ satisfies the forward Kolmogorov system of differential equations

$$
\frac{d}{dt}\boldsymbol{p}(t) = A(t)\boldsymbol{p}(t), \tag{6}
$$

where $A(t)$ is the transposed rate matrix i.e. $a_{ij}(t) = q_{ji}(t)$, $i, j \in \mathcal{X}$. Denote $\boldsymbol{f}(t) = (a_{10}(t), \ldots, a_{S0}(t))^{\mathrm{T}}$ and $\boldsymbol{z}(t) = (p_1(t), \ldots, p_S(t))^{\mathrm{T}}$ and introduce the new matrix[4] $B(t)$ of size $S \times S$, with the (i, j) entry $b_{ij}(t)$ equal to

$$
b_{ij}(t) = a_{ij}(t) - a_{i0}(t), \quad 1 \leq i, j \leq S.
$$

Using the normalization condition $p_0(t) = 1 - \sum_{i=1}^{S} p_i(t)$, the system (6) can rewritten as

$$
\frac{d}{dt}\boldsymbol{z}(t) = B(t)\boldsymbol{z}(t) + \boldsymbol{f}(t).
$$

All bounds of the rate of convergence to the limiting regime for $X(t)$ correspond to the same bounds of the solutions of the system

$$
\frac{d}{dt}\boldsymbol{y}(t) = B(t)\boldsymbol{y}(t), \tag{7}
$$

where $\boldsymbol{y}(t) = (y_1(t), \ldots, y_S(t))^{\mathrm{T}}$ is the vector with the elements of arbitrary signs (not necessarily all non-negative as in $\boldsymbol{p}(t)$). As it was firstly noticed in [18], it is more convenient to study the rate of convergence using the transformed version $B(t)$ given by $B^*(t) = TB(t)T^{-1}$, where T is the $S \times S$ upper triangular matrix of the form

---

[4] In other papers this matrix is sometimes called the reduced intensity matrix. It does not have any probabilistic interpretation.

$$T = \begin{pmatrix} 1 & 1 & 1 & \cdots & 1 \\ 0 & 1 & 1 & \cdots & 1 \\ 0 & 0 & 1 & \cdots & 1 \\ \vdots & \vdots & \vdots & \ddots & \vdots \\ 0 & 0 & 0 & \cdots & 1 \end{pmatrix}, \quad T^{-1} = \begin{pmatrix} 1 & -1 & 0 & \cdots & 0 \\ 0 & 1 & -1 & \cdots & 0 \\ 0 & 0 & 1 & \cdots & 0 \\ \vdots & \vdots & \vdots & \ddots & \vdots \\ 0 & 0 & 0 & \cdots & 1 \end{pmatrix}.$$

After some algebraic manipulations it can be seen that for the queueing system under consideration the matrix $B^*(t)$ is equal[5] to

$$B^*(t) = \begin{pmatrix} -(\lambda(t)+\mu_1(t)) & \mu_1(t) & 0 & \cdots & 0 & 0 & 0 \\ 0 & -(\lambda(t)+\mu_2(t)) & \mu_2(t) & \cdots & 0 & 0 & 0 \\ \lambda(t) & 0 & -(\lambda(t)+\mu_3(t)) & \cdots & 0 & 0 & 0 \\ \vdots & \vdots & \vdots & \ddots & \vdots & \vdots & \vdots \\ 0 & 0 & 0 & \cdots & 0 & -(\lambda(t)+\mu_{S-1}(t)) & \mu_{S-1}(t) \\ 0 & 0 & 0 & \cdots & \lambda(t) & -\lambda(t) & -\mu_S(t) \end{pmatrix}.$$

Introduce the new notation $u(t) = Ty(t)$. Then the system (7) can be rewritten in the form

$$\frac{d}{dt}u(t) = B^*(t)u(t), \tag{8}$$

where $u(t) = (u_1(t), \ldots, u_S(t))^T$ is, as well as $y(t)$, the vector with the elements of arbitrary signs (not necessarily all non-negative as in $p(t)$). Notice that one has converted the system (6), describing the probabilistic dynamic of the total number of customers in the considered queue, to the system (8), which looks the same as (1). So (8) is the starting point for the application of the proposed method.

Consider the case (ii), which implies that the service rates $\mu_i(t)$ in the matrix $B^*(t)$ are equal to $\mu_1(t) = 0$ and $\mu_i(t) = \mu(t)$, $2 \leq i \leq S$. The sequence of steps by which one applies the method depends on whether the arrival rate is "larger" or "smaller" than the service rate. At the expense of some loss of generality[6] only the "larger" case is considered below. Let $u(t)$ be the solution of (8). Remember that there are 2^S possible combinations of elements' signs in $u(t)$. Assume that all elements of the $u(t)$ are positive i.e. $u_i(t) > 0$, $1 \leq i \leq S$. Put $d_S = 1$, $d_{S-1} = \delta^{-1}$, $d_{S-2} = \delta$, and $d_{k-1} = \delta d_k$, for $1 \leq k \leq S - 2$, where $\delta > 1$. Denoting $w(t) = Du(t)$, where $D = diag(d_1, \ldots, d_S)$, (8) can be rewritten in the form

$$\frac{d}{dt}w(t) = B^{**}(t)w(t),$$

---

[5] Note that whenever the matrix $B^*(t)$ after all these transformations turns out to be essentially non-negative for any $t \geq 0$ i.e. $b_{ij}^*(t) \geq 0$ for $i \neq j$, the rate of convergence can be studied using the logarithmic norm method (see [24, 25]).

[6] Although the "smaller" case is not treated here, there is no principal difficulty, but longer sequence of steps in dealing with it. Note also that here the terms "larger" and "smaller" should be understood in the integral sense.

where $B^{**}(t) = DB^*(t)D^{-1}$. Let us write out the column sums of $B^{**}(t)$. For the sake of brevity introduce the notation $-\alpha_i(t) = \sum_{j=1}^{S} b_{ji}^{**}(t)$. Then

$$
\begin{aligned}
\alpha_1(t) &= \lambda(t) - \delta^{-2}\lambda(t), \\
\alpha_2(t) &= (\lambda(t) + \mu(t)) - \delta^{-2}\lambda(t), \\
\alpha_k(t) &= (\lambda(t) + \mu(t)) - \delta^{-2}\lambda(t) - \delta\mu(t), \ 3 \le k \le S - 3, \\
\alpha_{S-2}(t) &= (\lambda(t) + \mu(t)) - \delta^{-1}\lambda(t) - \delta\mu(t), \\
\alpha_{S-1}(t) &= (\lambda(t) + \mu(t)) + \delta\lambda(t) - \delta^2\mu(t), \\
\alpha_S(t) &= \mu(t) - \delta^{-1}\mu(t).
\end{aligned}
$$

Hence for this interval one can bound the corresponding $\alpha^I(t)$ by

$$
\alpha^I(t) = \min_{1 \le i \le S} \alpha_i(t) = \left(1 - \delta^{-1}\right) \min\left(\mu(t), \lambda(t) - \delta\mu(t)\right). \tag{9}
$$

The second argument in the $\min(\cdot, \cdot)$ function is positive since $\lambda(t)$ is assumed to be larger than $\mu(t)$. Assume now that $u_S(t) < 0$ and all other elements of $\boldsymbol{u}(t)$ are positive i.e. $u_i(t) > 0$, $1 \le i \le S - 1$. Put $d_S = -1$, $d_{S-1} = \delta$ and $d_{k-1} = \delta d_k$, for $1 \le k \le S - 1$, where $\delta > 1$. Then

$$
\begin{aligned}
\alpha_1(t) &= \lambda(t) - \delta^{-2}\lambda(t), \\
\alpha_2(t) &= (\lambda(t) + \mu(t)) - \delta^{-2}\lambda(t), \\
\alpha_k(t) &= (\lambda(t) + \mu(t)) - \delta^{-2}\lambda(t) - \delta\mu(t), \ 3 \le k \le S - 3, \\
\alpha_{S-2}(t) &= (\lambda(t) + \mu(t)) + \delta^{-2}\lambda(t) - \delta\mu(t), \\
\alpha_{S-1}(t) &= (\lambda(t) + \mu(t)) - \delta^{-1}\lambda(t) - \delta\mu(t), \\
\alpha_S(t) &= \mu(t) + \delta\mu(t).
\end{aligned}
$$

Hence for this interval one can bound the corresponding $\alpha^I(t)$ by

$$
\alpha^I(t) = \min_{1 \le i \le S} \alpha_i(t) = \left(1 - \delta^{-1}\right)\left(\lambda(t) - \delta\mu(t)\right). \tag{10}
$$

Moreover one can note that in all other $2^S - 2$ cases only negative elements in the columns of the matrix $B^*(t)$ can be added. Thus in all other intervals the values of $\alpha^I(t)$ is greater for the same $|d_k|$. Thus one obtains the following upper bound for the rate of convergence for the queueing system (ii):

$$
\|\boldsymbol{u}(t)\| \le C^* e^{-\int_0^t \alpha^*(u)\,du} \|\boldsymbol{w}(0)\|, \tag{11}
$$

for any $t \ge 0$, where $C^* = \delta^S$, $\alpha^*(t) = \left(1 - \delta^{-1}\right)\min\left(\mu(t), \lambda(t) - \delta\mu(t)\right)$. Moreover $X(t)$ is weakly ergodic and the following bound on the rate of convergence holds:

$$
\|\boldsymbol{p}^*(t) - \boldsymbol{p}^{**}(t)\| \le 4C^* e^{-\int_0^t \alpha^*(u)\,du} \|\boldsymbol{w}(0)\|, \tag{12}
$$

for any initial conditions.

Even though the case (i) can be treated in the same way as described above, it is more convenient to treat it differently. Notice that in the case (i) all off-diagonal elements of the matrix $B^*(t)$ are non-negative and the sums $\sum_{j=1}^{S} b_{ji}^*(t)$

for the matrix $B^*(t)$ are equal to $-\mu(t)$. Thus the logarithmic norm of the matrix $B^*(t)$ is $\gamma(B^*(t)) = -\mu(t)$ and one can apply the approach based on the notion of the logarithmic norm. The results from the papers [6, 19, 25] immediately give that $X(t)$ is weakly ergodic and the following bounds on the rate of convergence hold:

$$\|u(t)\| \le e^{-\int_0^t \mu(\tau)\, d\tau} \|u(0)\|, \tag{13}$$

$$\|p^*(t) - p^{**}(t)\| \le 4e^{-\int_0^t \mu(\tau)\, d\tau} \|u(0)\|, \tag{14}$$

for any initial conditions.

4   Numerical Example

Using the proposed method one can calculate not only the rate of convergence but also the approximate values for the limiting performance characteristics of the process $X(t)$ for appropriate interval $[t_1, t_2]$ with the known approximation error (see steps (a)—(d) in the Sect. 1.)

Let in the queue considered in Sect. 3 the functions $\lambda(t)$ and $\mu(t)$ be 1−periodic functions equal to $\lambda(t) = 4 + \sin(2\pi t)$ and $\mu(t) = 1 + \cos(2\pi t)$, respectively[7]. Let $S = 100$. Then by applying the convergence bounds[8] of the

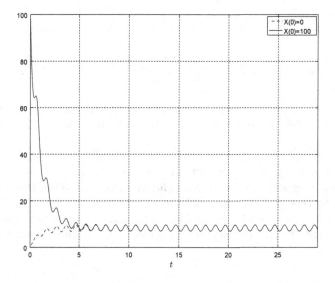

Fig. 1. Case (i). Rate of convergence of the mean number of customers in the system in the interval $[0, 30]$ for two different initial system occupancies ($X(0) = 0$ and $X(0) = 100$).

---

[7] Such choice of functions is justified as follows. Firstly, the results in Sect. 3 are presented for the case when $\lambda(t)$ is "larger" than $\mu(t)$. Secondly for 1−periodic functions it is easier to decide which regime is reasonable to regard as a limiting one (see also the Footnote 2).

[8] For the case (i) the bound (14), for the case (ii) the bound (12).

previous section, one can compute, for example, the limiting value of the mean number of customers in the systems i.e. $\sum_{i=0}^{S} i p_i(t)$. For the case (i) in Fig. 1 one can see two graphs of the mean number of customers in the system at time t corresponding to two different initial conditions: when initially the system is empty (lower graph) and when initially the system is full (upper graph). The graphs are getting closer to each other as time t increases and eventually both coincide with the "limiting" graph, depicted in Fig. 2.

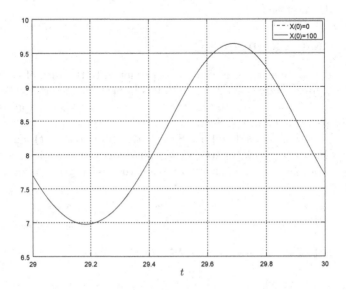

Fig. 2. Case (i). The limiting mean number of customers in the system in the interval $[29, 30]$ for two different initial system occupancies ($X(0) = 0$ and $X(0) = 100$).

Figures 3 and 4 provide the same information for the mean number of customers in the system for the case (ii). The time interval $[0, 30]$ (for both cases) was chosen by repeated attempts, shifting the right end of the interval until the convergence has become clearly visible. Note that by comparing Figs. 1 and 3 one can see that the convergence rate in the case (ii) is much slower than in the case (i).

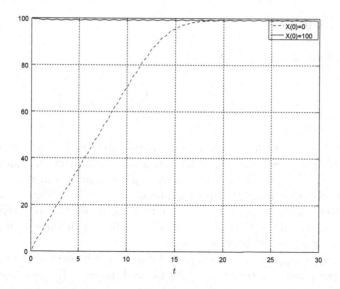

Fig. 3. Case (ii). Rate of convergence of the mean number of customers in the system in the interval $[0, 30]$ for two different initial system occupancies ($X(0) = 0$ and $X(0) = 100$).

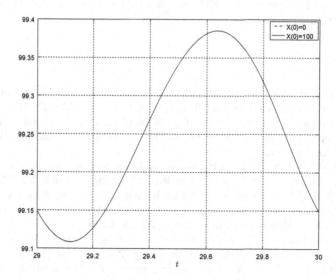

Fig. 4. Case (ii). The limiting mean number of customers in the system in the interval $[29, 30]$ for two different initial system occupancies ($X(0) = 0$ and $X(0) = 100$).

5   Conclusion

Coming back to (3) one can note that

$$\alpha^I(t) \leq \max_{1 \leq j \leq S} \left(k_{jj}(t) + \sum_{i=1, i \neq j}^{S} \frac{d_i^I}{d_j^I} |k_{ij}(t)| \right).$$

By putting $d_i^I = 1$ for all $1 \leq i \leq S$, one immediately arrives at the inequality $\alpha(t) \leq \gamma(K(t))$, where $\gamma(K(t))$ is the logarithmic norm of the matrix $K(t)$. Thus the method proposed in Sect. 2 always gives results, which are no worse than results obtained using the approach based one the logarithmic norm. Since the logarithmic norm method gives exact bounds in the case of essential non-negativity of the matrix $K(t)$ (see [22]), the method of differential inequalities yields exact estimates in this case as well.

The proposed approach can be applied if and only if there is an opportunity to find proper constants $\{d_i^I, 1 \leq i \leq S\}$ for each interval I. Since (apparently) there does not exist any general algorithm for selecting $\{d_i^I, 1 \leq i \leq S\}$ for a general inhomogeneous birth and death process with a finite state space, the scope of the proposed approach is hard to define. For every new problem instance one has to examine the matrix $K(t)$ and act on the trial and error basis, when searching for $\{d_i^I, 1 \leq i \leq S\}$.

References

1. Andersen, A.R., Nielsen, B.F., Reinhardt, L.B., Stidsen, T.R.: Staff optimization for time-dependent acute patient flow. Eur. J. Oper. Res. **272**(1), 94–105 (2019)
2. van Brummelen, S.P.J., de Kort, W.L., van Dijk, N.M.: Queue length computation of time-dependent queueing networks and its application to blood collection. Oper. Res. Health Care **17**, 4–15 (2018)
3. Chen, A.Y., Pollett, P., Li, J.P., Zhang, H.J.: Markovian bulk-arrival and bulk-service queues with state-dependent control. Queueing Syst. **64**, 267–304 (2010)
4. Di Crescenzo, A., Giorno, V., Krishna Kumar, B., Nobile, A.G.: A time-non-homogeneous double-ended queue with failures and repairs and its continuous approximation. Mathematics **6**(5) (2018). Article ID 81
5. Giorno, V., Nobile, A.G., Spina, S.: On some time non-homogeneous queueing systems with catastrophes. Appl. Math. Comp. **245**, 220–234 (2014)
6. Granovsky, B., Zeifman, A.: Nonstationary queues: estimation of the rate of convergence. Queueing Syst. **46**, 363–388 (2004)
7. Kartashov, N.V.: Criteria for uniform ergodicity and strong stability of Markov chains with a common phase space. Theory Probab. Appl. **30**, 71–89 (1985)
8. Liu, Y.: Perturbation Bounds for the stationary distributions of Markov chains. SIAM J. Matrix Anal. Appl. **33**(4), 1057–1074 (2012)
9. Meyn, S.P., Tweedie, R.L.: Stability of Markovian processes III: foster- Lyapunov criteria for continuous time processes. Adv. Appl. Probab. **25**, 518–548 (1993)
10. Mitrophanov, A.Y.: Stability and exponential convergence of continuous-time Markov chains. J. Appl. Probab. **40**, 970–979 (2003)

11. Mitrophanov, A.Y.: The spectral gap and perturbation bounds for reversible continuous-time Markov chains. J. Appl. Probab. **41**, 1219–1222 (2004)
12. Mitrophanov A.Y.: Connection between the rate of convergence to stationarity and stability to perturbations for stochastic and deterministic systems. In: Proceedings of the 38th International Conference Dynamics Days Europe, DDE 2018, Loughborough, UK (2018). http://alexmitr.com/talk_DDE2018_Mitrophanov_FIN_post_sm.pdf
13. Rudolf, D., Schweizer, N.: Perturbation theory for Markov chains via Wasserstein distance. Bernoulli **24**(4A), 2610–2639 (2018)
14. Satin, Y., Zeifman, A., Kryukova, A.: On the rate of convergence and limiting characteristics for a nonstationary queueing model. Mathematics **7**(8), 678 (2019)
15. Schwarz, J.A., Selinka, G., Stolletz, R.: Performance analysis of time-dependent queueing systems: survey and classification. Omega **63**, 170–189 (2016)
16. Tan, X., Knessl, C., Yang, Y.: On finite capacity queues with time dependent arrival rates. Stoch. Process. Appl. **123**(6), 2175–2227 (2013)
17. Zeifman, A.I.: Stability for continuous-time nonhomogeneous Markov chains. In: Kalashnikov, V.V., Zolotarev, V.M. (eds.) Stability Problems for Stochastic Models. LNM, vol. 1155, pp. 401–414. Springer, Heidelberg (1985). https://doi.org/10.1007/BFb0074830
18. Zeifman, A.I.: Properties of a system with losses in the case of variable rates. Autom. Remote Control **50**(1), 82–87 (1989)
19. Zeifman, A., Leorato, S., Orsingher, E., Satin, Ya., Shilova, G.: Some universal limits for nonhomogeneous birth and death processes. Queueing Syst. **52**, 139–151 (2006)
20. Zeifman, A.I., Korolev, V.Y.: On perturbation bounds for continuous-time Markov chains. Stat. Probab. Lett. **88**, 66–72 (2014)
21. Zeifman, A., Korotysheva, A., Korolev, V., Satin, Y., Bening, V.: Perturbation bounds and truncations for a class of Markovian queues. Queueing Syst. **76**, 205–221 (2014)
22. Zeifman, A.I., Korolev, V.Y.: Two-sided bounds on the rate of convergence for continuous-time finite inhomogeneous Markov chains. Stat. Probab. Lett. **103**, 30–36 (2015)
23. Zeifman, A.I., Korotysheva, A.V., Korolev, V.Y., Satin, Y.A.: Truncation bounds for approximations of inhomogeneous continuous-time Markov chains. Theory Probab. Appl. **61**(3), 513–520 (2017)
24. Zeifman, A., et al.: On sharp bounds on the rate of convergence for finite continuous-time Markovian Queueing models. In: Moreno-Diaz, R., Pichler, F., Quesada-Arencibia, A. (eds.) Computer Aided Systems Theory EUROCAST 2017. LNCS, vol. 10672, pp. 20–28. Springer, Cham (2018)
25. Zeifman, A., Razumchik, R., Satin, Y., Kiseleva, K., Korotysheva, A., Korolev, V.: Bounds on the rate of convergence for one class of inhomogeneous Markovian queueing models with possible batch arrivals and services. Int. J. Appl. Math. Comput. Sci. **28**, 141–154 (2018)
26. Zeifman, A., Satin, Y., Kiseleva, K., Korolev, V., Panfilova, T.: On limiting characteristics for a non-stationary two-processor heterogeneous system. Appl. Math. Comput. **351**, 48–65 (2019)
27. Zeifman, A., Satin, Y., Kiseleva, K., Kryukova, A.: Applications of differential inequalities to bounding the rate of convergence for continuous-time Markov chains. In: AIP Conference Proceedings, vol. 2116, Article ID 090009 (2019)

Author Index

Alhawas, Albatool 1

Baier, Christel 133
Bernardo, Marco 16
Boon, Marko 65

Camilli, Matteo 33
Capra, Lorenzo 33

Dubslaff, Clemens 133

Eijkelenboom, Gerard 65
Ezhilchelvan, Paul 50

Gaeta, Rossano 84

Klaasse, Bo 65
Klüppelholz, Sascha 133
Knottenbelt, William J. 100
Kryukova, Anastasia 148
Kumar, Akash 133

Liquori, Luigi 84

Marchenko, Yuriy 100
Marin, Andrea 118
Mitrani, Isi 50

Razumchik, Rostislav 148
Rossi, Sabina 118

Sardar, Muhammad Usama 133
Satin, Yacov 148
Sereno, Matteo 84
Shilova, Galina 148

Thomas, Nigel 1
Timmerman, Rik 65

van Ballegooijen, Tessel 65

Waudby, Jack 50
Webber, Jim 50
Wolter, Katinka 100

Zeifman, Alexander 148

Printed in the United States
By Bookmasters